TURBINE THERMAL
APPRAISAL

TURBINE THERMAL APPRAISAL

A Spreadsheet Approach

Book I. Nozzles

Ernesto Novillo

Library of Congress Control Number: 2016902305
ISBN: Hardcover   978-1-5144-5992-8
                 Softcover   978-1-5144-5991-1
                 eBook   978-1-5144-7043-5

Print information available on the last page.

Rev. date: 03/04/2016

To order additional copies of this book, contact:
Xlibris
1-888-795-4274
www.Xlibris.com
Orders@Xlibris.com
735935

CONTENTS

Objective and Philosophy of this Book... xv

Symbols, Nomenclature and Units xix

PART I. NOZZLE OPERATION

Chapter 1. Phenomena Inside a Nozzle.. 3

    1.1. Presentation of nozzle and diffuser technology 4

        1.1.1. Today's technology 5

    1.2. The basic physical laws in a nozzle 8

        1.2.1. The principle of conservation of energy 9

        1.2.2. The ideal gas equation of state......................11

        1.2.3. The principle of flow mass continuity.............12

        1.2.4. Isentropic relations 13

        1.2.5. Enthalpy and internal energy formulas..........15

    1.3. Energy conversion in nozzle flow 16

    1.4. Flow velocity behavior in nozzles.............................19

    1.5. Flow critical properties 20

    1.6. Discharge jet shape 20

    1.7. Specific mass flow...20

    1.8. Turbulence in nozzles 22

    1.9. Shock waves in CD nozzles......................................24

    1.10. Flow choking 26

    1.11. Example of a critical ratio calculation 29

Chapter 2. Available energy and efficiency of nozzles..............32

    2.1. Available energy for $V_1 = 0$ 32

2.2. Energy efficiency of a nozzle................................35

2.3. Friction losses and enthalpy discharge 36

2.5. Total available energy and discharge enthalpy
for $V_1 \neq 0$..38

2.6. Steam expansion in a nozzle...............................39

Chapter 3. Velocity and mass flow in nozzles...............42

3.1. Flow velocity calculation with enthalpy drop............42

3.2. Saint Venant - Wantzel equation for ideal gases........44

3.3. Speed of sound and critical ratios.......................45

3.4. Example of critical properties calculation..................47

3.5. Throat and outlet velocities................................48

3.6. Example of velocity curves calculation.....................49

3.7. Mass flow in a choked flow 51

3.8. Example of a critical velocity and mass flow
calculation 55

3.8.1. Case a) p_2 = 4 kg / cm2..............................56

3.8.2. Case b) p_2 = 14 kg/cm2 a 59

Chapter 4. Flow perturbations: Friction, Turbulence
and Shock Waves..62

4.1. Friction in nozzles..62

4.1.1. The Moody diagram 65

4.1.2. Velocity coefficient formula............................67

4.2. Example of nozzle efficiency sensitivity versus
relative roughness..69

4.2. Example of nozzle efficiency calculation................71

4.4. Example of turbulence assessment in
conical nozzle...73

4.5. Normal shock waves in CD nozzles 78

4.6. Shock wave identification...83

4.7. Example of shock wave calculation 88

4.8. Example of a turbine power affected by a normal shock wave 91

Chapter 5. Geometric design and performance curves.............93

5.1. Cross section area calculation 93

5.1.1. Formula of area versus pressure ratio r_{px} for ideal gas...94

5.1.2. Example of nozzle profile and cross sections area calculation...................................96

5.1.3. Formula of area versus M_x 99

5.1.4. Formula of inlet area for CD conical nozzles 100

5.2. Flow shape and optimum design condition............ 102

5.3. Expansion ratio definitions and types of expansion.. 104

5.3.2. Overexpansion. $X_g > X_f$................................ 106

5.3.3. Underexpansion. $X_g < X_f$............................. 106

5.3.4. Notes to expansion types............................ 107

5.4. Velocity coefficient. US Naval Institute 108

5.5. Velocity coefficient. C. P. Steinmetz curve................ 111

5.6. Geometric design procedure 114

5.7. Example of a CD nozzle design 115

5.7.1. Technical specifications............................... 115

5.7.2. Expansion ratios 123

5.7.3. Nozzle performance curves 124

5.8. Summary of most important nozzle aspects............ 128

PART II. NOZZLE PROJECTS

Chapter 6. Gas Nozzle Project 133

    6.1. Project plan 133

    6.2. Input data and physical constants 134

    6.3. Critical properties and nozzle type identification 137

    6.4. Thermodynamics properties 138

    6.5. Energy properties 139

    6.6. Flow properties 141

    6.7. Geometric design 143

    6.8. Expansion ratios 145

    6.9. Performance table and curves 147

Chapter 7. Steam nozzle project 149

    7.1. Project technical specifications 151

    7.2. Case 2. Critical properties and nozzle type
        identification 154

    7.3. Case 2. Thermodynamic properties 154

    7.4. Case 2. Energy properties 156

    7.5. Case 2. Geometric design of conical nozzle 158

    7.6. Case 2. Flow properties 159

    7.7. Case 2. Expansion ratios and efficiency validation 159

    7.8. Case 2. Performance curves 160

    7.9. Efficiency assessment under operating
        conditions other than design specifications 162

        7.9.1. Case 1. Efficiency assessment for
            $p_2 = 1.2$ kg/cm2 a 162

        7.9.2. Case 5. Efficiency assessment for
            $p_2 = 5.0$ kg/cm^2 164

7.10. Cases 1 to 7. Nozzle operation sensitivity to discharge pressure..167

    7.10.1. Power sensitivity to discharge pressure......167

    7.10.2. Flow cross section areas sensitivity to discharge pressure 169

    7.10.3. Velocity coefficient and efficiency sensitivity to discharge pressure 170

7.11. Example of normal shock waves calculation.........172

Chapter 8. Derivation of formulas...................................177

    8.1. Alternative formula for the available energy of gases...177

    8.2. Conversion efficiency formula..............................178

    8.3. Saint Venant - Wantzel equation.........................179

    8.4. Critical ratios formulas 181

    8.5. Mass flow versus nozzle pressure ratio...............184

    8.6. Thermodynamic properties versus Mach number......186

    8.7. Cross section area versus pressure ratio r_{px}............189

    8.8. Cross section area versus Mach number...............191

Bibliography...195

List of Figures...199

List of Uploaded Files..203

Index...205

This book must be read along with the
spreadsheet files uploaded in:

https://www.facebook.com/groups/turbinia

Spreadsheet files are free to download
and have no property copyright

Dedicated to Marité... my life support
and to my son Ernie, the meritorious editor of this book
and to my daughter Monica, the artist who designed the cover

Objective and Philosophy
of this Book

This is a book of practical Thermotechnics for turbines, aimed at technicians and engineers who have responsibilities in turbine facilities, or participate in engineering projects where turbines are a component of a larger complex. This book is also aimed at professionals who are responsible for turbine procurement, either in its complete form or for acquisition of spare parts.

The different topics have been developed on the basis of Thermodynamics and Fluid Mechanics and conceived as the application of these sciences to nozzle engineering. Therefore, no theoretical discussions are included in the book, just the technical application of formulas and methods, some of them developed by the author. The theoretical basis can be consulted in the excellent and extensive bibliography about Thermodynamics and Fluid Mechanics. Some specialized information and recommended texts are under the Bibliography title at the end of the book.

Part I (Chapters 1 to 5) describes the process that takes place in the nozzle. Useful formulas of practical interest are presented to implement the nozzle design and performance assessment. These formulas are used in Part II, which is only dedicated to the project of both a gas nozzle (Chapter 6) and a steam nozzle (Chapter 7). The steam nozzle project culminates with the impact assessment of operating conditions other than design conditions, which are mainly due to the generation of turbulence and shock waves.

In both Parts I and II mathematical derivations have been avoided to focus only on final practical formulas and how to use them to design or assess nozzle behavior. No calculus operations have been included in this book, only algebraic expressions.

However, most important formulas discussed in the text are derived in Chapter 8. This chapter isn't essential for the purpose of the book. It's only useful to those professionals interested in seeing the mathematical fundaments of formulas set forth in the text.

An important part of this book's philosophy is to explain easy-to-use mathematical tools to evaluate actual operational turbines, or generate behavior models of a turbine operation for benchmarking purposes. These mathematical tools are specifically developed in spreadsheets to solve practical problems. These tools can be downloaded free from https://www.facebook.com/groups/turbinia. It's recommended to download all spreadsheet files, for these encompass almost 40% of the book.

To facilitate the application of spreadsheets, the use of charts and graphs has been avoided in the calculation procedures. Instead, all calculations have been designed for their implementation in spreadsheets. Conceptual explanations have been reinforced by means of behavior curves, which are substantial only for a better understanding of the topic under study. However, it's strongly recommended to use the Mollier chart in the design and assessment of steam nozzles, due to its clear graphical representation of steam transformations.

All spreadsheet calculations are based on theoretical formulas, valid for practical applications, and other empirical formulas, to avoid the use of tables or graphs.

The use of spreadsheet as a mathematical tool for design or assessment evaluations must be considered a "theoretical and acceptable" approach, according to formulas of Thermodynamic and Mechanic of Fluids. However, turbulence is a complex to simulate phenomenon and results of algebraic formulas barely assess the turbulence impact on nozzle performance. Fortunately, significant advances in Computational Fluid Dynamics (CFD), which calculate trajectories of hundreds of thousands of fluid particles, is a highly recommended technology, mainly to optimize the nozzle design. Different software packages applicable to turbomachinery are available in the market and some of them are open source.

Formulas have a single system of units, which is the SI. All formulas have consistent units, hence, they don't require transformations of any kind, except pressures and nozzle dimensions, which use centimeters instead of meters.

I'd like to mention the merit of numerous people in teaching me the theory and practice of turbine technologies and their supporting science. To all of them I am very grateful for the teachings I received in the ships of the Argentine Navy, the National University of Cordoba in Argentina, and in different countries where I worked on projects for the installation, maintenance and commissioning of turbines, mainly for naval propulsion and power generation. I cannot list the names of all these people who helped me so much over more than 40 years of professional activity, but I would like to manifest my remembrance and tribute to all of them.

Ernesto Novillo
Electrical and Mechanical Engineer
P.Eng, Canada
P.Energy Mgr, USA
Canada, 3rd of December, 2015

Symbols, Nomenclature and Units

• Latin letters

a_V: Velocity constant, $|m/s|/|\sqrt{°K}|$

a_G: Mass flow constant, $\sqrt{m}\,/s$

A_1: Inlet section area of the nozzle, m^2. Also expressed in cm^2

A_2: Outlet section area of the nozzle, m^2. Also expressed in cm^2

A_{ft}: Flow area at the throat, m^2. Also expressed in cm^2

$A_f(x)$: Section area of the flow at a distance x of the inlet, m^2

$A_n(x)$: Section area of the nozzle at a distance x of the inlet, m^2

A_t: Throat area of the nozzle, m^2. Also expressed in cm^2

C: Convergent nozzle

CD: Convergent divergent nozzle

c_p: Specific heat at constant pressure, $kJ/|kg.°K|$

c_v: Specific heat at constant volume, $kJ/|kg.°K|$

D_1: Nozzle diameter at the inlet, m. Also expressed in cm

D_2: Nozzle diameter at the outlet, m. Also expressed in cm

D_x: Nozzle diameter at a distance x from the inlet, m. Also expressed in cm

D_t: Nozzle diameter at the throat, m. Also expressed in cm

D_{avg}: Nozzle average diameter, m. Also expressed in cm

D_{eq}: Weighted nozzle diameter, m or cm

e_c: Conversion efficiency, p.u.

e_n: Nozzle efficiency, p.u.

e_t: Turbine efficiency, p.u.

E_a: Available energy, kJ/kg

E_k: Kinetic energy, kJ/kg

E_t: Total energy, kJ/kg

f: Friction factor, non-dimensional

F_{v1}: Volume flow in the convergent part of the nozzle, m^3/s

F_{v2}: Volume flow in the divergent part of the nozzle, m^3/s

g: Gravity acceleration = 9.81 m/s^2

g_{s1}: Specific mass flow at the inlet, kg/|m^2.s|. Also expressed in kg/|cm^2.s|

g_{st}: Specific mass flow at the throat, kg/|m^2.s|. Also expressed in kg/|cm^2.s|

g_{s2}: Specific mass flow at the outlet, kg/|m^2.s|. Also expressed in kg/|cm^2.s|

g_{sx}: Specific mass flow at a distance x from the inlet, kg/|m^2.s|. Also expressed in kg/|cm^2.s|

G: Mass flow, kg/s

G_m: Maximum mass flow under choked conditions, kg/s

GT: Gas turbine

h_x: Enthalpy at a distance x from the inlet, kJ/|kg.°K|

h_1: Enthalpy at the inlet, kJ/|kg.°K|

h_2: Enthalpy at the outlet, kJ/|kg.°K|

h_e: Latent heat or evaporation heat, kJ/|kg.°K|

h_l: Sensible heat, kJ/|kg.°K|

h_s: Total enthalpy of saturated steam, kJ/|kg.°K|

h_t: Enthalpy at the throat, kJ/|kg.°K|

J: Mechanical equivalent of heat. = 102 kJ/|kg.m|

k: Specific heat ratio, c_p/c_v

K_a: Area constant of the nozzle, cm^2/|kg.s|

L_1: Length of the convergent part of the nozzle, m. Also expressed in cm

L_2: Length of the divergent part of the nozzle, m. Also expressed in cm

L_t: Total length of the nozzle, m. Also expressed in cm

M: Mach number

M_1: Mach number at the nozzle inlet, non-dimensional

M_2: Mach number at the nozzle outlet, non-dimensional

M_t: Mach number at the nozzle throat, non-dimensional

M_y: Mach number at a distance x of the nozzle inlet. Also used for Mach number at a shock wave inlet

MW: Molecular weight. Expressed in |kg/kmol|.

M_z: Mach number at a shock wave inlet, non-dimensional

N: Nozzles quantity in a turbine, non-dimensional

p_1: Pressure at the nozzle inlet, kg/cm² a

p_2: Pressure at the nozzle outlet, kg/cm² a

p_A: Discharge pressure in a nozzle with inner shock waves, kg/cm² a

p_B: Discharge pressure at the shock wave pressure curve, kg/cm² a

p_c: Critical pressure, kg/cm²

p_D: Design pressure, kg/cm$_2$ a

p_s: Saturation pressure, kg/cm²

p_t: Pressure at the throat, kg/cm² a

p_x: Pressure at a distance x from the inlet

p_y. Pressure at a shock wave inlet, kg/cm² a

p_z: Pressure at a shock wave outlet, kg/cm² a

MW: Molecular weight

q_f: Friction heat, kJ/|kg|

q_n: Net heat interchanged between the system and the environment, kJ/kg

P: Turbine power, kJ/s or kW

R: Specific gas constant, kJ/|kg.°K|

R_U: Universal gas constant = 8.315 kJ/|kmol.°K|

R_e: Reynolds number, non-dimensional

r_{pc}: Critical pressure ratio, p_{tc}/p_1, non-dimensional. For choked flow

r_{pn}: Nozzle pressure ratio p_2/p_1

r_{po}: Pressure ratio of the nozzle at the minimum area, p_o/p_1, non-dimensional

r_{pt}: Pressure ratio at the throat, p_t/p_1, non-dimensional For no choked flow.

r_{px}: Pressure ratio of the nozzle at a distance x of the inlet, p_x/p_1, non-dimensional

r_{tc}: Critical temperature ratio, non-dimensional

r_{vc}: Critical specific volume ratio, non-dimensional

r_A: Ratio of the geometric and flow cross sections area

S_1: Initial entropy of a transformation, kJ/|kg.°K|

S_2: Final entropy of a transformation, kJ/|kg.°K|

S_u: Sutherland constant, °K

S_y: Entropy at a shock wave inlet, kJ/|kg.°K|

t_c: Total circulation time of the nozzle, ms

t_{c1}: Circulation time in the convergent part of the nozzle, ms

t_{c2}: Circulation time in the divergent part of the nozzle, ms

T_1: Temperature at the inlet, °K

T_2: Temperature at the outlet, °K

T_c: Critical temperature, °K

T_s: Saturation temperature, °K

T_t: Temperature at the throat, °K

T_y: Temperature at a distance x from the inlet. Also used for temperature at a shock wave inlet, °K

Tz: Temperature at a shock wave outlet, °K

u: Internal energy, kJ/kg

v_1: Specific volume at the nozzle inlet, m³/kg

v_2: Specific volume at the nozzle inlet, m³/kg

v_c: Critical specific volume at the throat, m³/kg

v_t: Specific volume at the throat, m³/kg

v_y: Specific volume at the shock wave inlet, m³/kg

v_z: Specific volume at the shock wave outlet, m³/kg

V_1: Gas or steam velocity at the inlet, m/s

V_2: Gas or steam velocity at the outlet, m/s

Vo: Output flow velocity from an orifice, m/s

V_s: Speed of sound, m/s

V_{s1}: Speed of sound at the inlet, m/s

V_{s2}: Speed of sound at the outlet, m/s

V_t: Gas or steam velocity at the throat, m/s

V_x: Gas or steam velocity at a distance x from the inlet, m/s

Vol$_1$: Volume of the convergent part of the nozzle, m³

Vol$_2$: Volume of the divergent part of the nozzle, m³

Vol$_n$: Total nozzle volume, m³

W_1: Gas or steam weight in the convergent part of the nozzle, kg

W_2: Gas or steam weight in the divergent part of the nozzle, kg

W_e: Weight of dry steam (evaporated) into the mixture, kg

W_l: Weight of liquid water into the mixture, kg

Wn: Net work interchanged between the system and the environment, kJ/kg

x: Steam title or distance to the inlet section, non-dimensional

X_g: Geometric expansion ratio, A_2/A_t, non-dimensional

X_f: Flow expansion ratio or theoretical expansion ratio, A_{2f}/A_{tf}, non-dimensional

X_v: Specific volume expansion ratio, v_2/v_t, non-dimensional

$X_{Velocity}$: Velocity expansion ratio, V_2/V_t, non-dimensional

• Greek letters

α_1: Convergent angle of the nozzle, °

α_2: Divergent angle of the nozzle, °

δ: Gas or steam specific gravity, kg/m³

ε: Inner wall roughness, m or cm

ε_r: Relative roughness, non-dimensional

μ: Dynamic viscosity, mass |kg/m.s|

ρ: Gas or steam density, mass kg/m³

φ:Velocity coefficient, $V_2/V_{2\,ideal}$, non-dimensional

ΔS: Entropy change, kJ/|kg.°K|

Δp_{cc}: Pressure loss in the combustion chamber of a gas turbine, kg/cm²

• Subscripts

1: Nozzle inlet or convergent part of the nozzle

2: Nozzle outlet or divergent part of the nozzle

actual: Property under friction conditions

air: A property of the air

B: Shock wave pressure at the nozzle outlet

c: Critical

D: Design pressure at the nozzle outlet

f: Flow

g: Geometric

ideal: Property under no friction conditions
o: Minimum nozzle area (orifice of Saint Venant equation)
s: Specific
t: Nozzle throat
x: Denotes a property that changes along the nozzle axis
y: shock wave inlet
z: shock wave outlet

PART I
NOZZLE OPERATION

CHAPTER 1

PHENOMENA INSIDE A NOZZLE

This chapter explains the conceptual operation of a nozzle. To do this the number of mathematical expressions used is minimized, because its objective is just the description of the flow behavior within a nozzle and not its calculation.

Turbine nozzles are aimed to convert fluid thermal energy into kinetic energy. They are short and relatively small devices. Their walls are strong enough to withstand the internal pressure to which they are subjected. Transformations inside nozzles can be studied as isentropic, however these transformations are not ideal because there exists friction between the fluid and inner walls. This friction releases heat and the aftermath is a loss of kinetic energy, which is significant and should be taken into account in calculating the design or performance evaluation.

The study of nozzles is supported by two sciences: Thermodynamics and Fluid Mechanics. However, the former has a smaller extensive application, because the flow regime is assumed to be one-dimensional. Actually it's also compressible and viscous. In a one-dimensional flow, the distribution of thermodynamic properties inside the nozzle is explained with only one spatial coordinate: the abscissa x (see Figure 1.1). Consequently, the flow properties are constant throughout its cross section. The adoption of one-dimensional flow is a convenient and very common assumption for the study of nozzles. This is not an exact assumption, but it's generally considered a satisfactory approximation in practice, at site installations.

Note: by property distribution it is understood a plot of a property represented as a curve, or table, versus the abscissa x.

1.1. Presentation of nozzle and diffuser technology

Technology, which is the application of science, has greatly succeeded in the Thermotechnics and Thermodynamic brotherhood. The most resonant example of this success are the heat engines, where the theoretical and practical principles of Thermodynamics and Thermotechnics, have proved to be extremely useful to develop one the most important engineering applications.

The history of this success began in 1697, when the military engineer Thomas Savery built the first steam engine to drive a pump for lifting water in a coal mine. Today it is well known that Savery's machine was very inefficient. However, other attempts to build a steam engine followed in the eighteenth century. But neither Savery nor his successors had a physical doctrine body that would allow them to analyze the theoretical basis of what they had built and used (and patented!) to improve their inventions. Consequently, Thermotechnics was initially developed without a strong scientific basis. Among many pioneers, the engineer James Watt deserves a special mention for the improvements he made to the initial steam engines. Watt's developments drove the Industrial Age in Europe and North America. His first machines, used to drive water pumps, were installed in 1775. It had been a century since the inefficient, but meritorious invention of Savery! The following important invention was made one century later to James Watt by Sir Charles Parsons on 1884, who successfully invented and fabricated the first steam turbine.

But the theoretical basis to explain the steam engine, was not known until 1824, when the military engineer Sadi Carnot published Reflexions sur la puissance motrice du feu propres et sur les machines à cette puissance developer (Reflections about the motive power of fire and power developer machines). With this book Carnot created Thermodynamics science. Thereafter the development of Thermodynamics and Thermotechnics grew extraordinarily. The first fed Thermotechnics with new ideas, which helped to design and improve fabrication processes and test procedures that led to the creation of modern steam engines, where turbines have a very important role.

1.1.1. Today's technology

It is common the use of fluids flowing along a pipe, or through any type of engine (turbine for example), where the flow's kinetic energy or pressure, requires to be changed. Some of the best known cases are in power generation, naval propulsion, industrial machines and aircraft engines.

Fortunately, it is possible to change fluid's properties by using very simple devices, with no moving parts, whose operation is described in this chapter. They are the very well-known nozzles and diffusers. Nozzles increase the kinetic energy of the flow by means of throttling it and diffusers reduce the fluid velocity and increase its pressure, by expanding the fluid in an increasing diameter pipe.

The nozzles are not designed to carry fluids or to produce work, but to expand a fluid to increase its kinetic energy. So, this augmented energy is transferred to a turbine drive wheel, as soon as the flow impacts the turbine rotor blades. The desirable flow operating regime within the nozzle is with minimum turbulence. This turbulence causes a significant reduction in the nozzle efficiency, which is manifested in the reduction of the discharge kinetic energy and the increase of the fluid enthalpy.

There are two geometrical types of nozzles: convergent divergent (CD) and convergent (C). This distinction does not exist in diffusers, since they are all divergent. Figure 1.1 shows the basic geometric shapes of the CD nozzle and the nomenclature applicable in the rest of this book. The angles α_1 and α_2 are called angle of convergence and angle of divergence respectively. Their typical ranges are 30° to 70° for α_1 and 5° to 15° for α_2. This announces that hereinafter, the inlet side is identified with the subscript 1, the throat with subscript t, and the discharge side with subscript 2. See nomenclature used for subscripts in section Symbols, Nomenclature and Units.

Since it is assumed that the flow is one-dimensional, the location of any physical property is determined by its distance to the nozzle inlet (abscissa x). From the physical standpoint, the one-dimensional flow assumption means that any physical property of the flow is constant throughout any cross section, either the nozzle or diffuser. However, this assumption is not correct due to the flow turbulence. Therefore, the correct term would be average property along a cross section instead of property.

Actually, nozzles are more rounded than the scheme shown in Figures Figure 1.1,1.2 and 1.3 and their straight section is not always circular, such as nozzles of turbine blade wheels. Standard values of nozzle proportions are found in publications of various professional institutions like ASME, ISO, etc. The convergent nozzle is simpler than the CD nozzle, because it is composed of only one body, as shown in Figure 1.2. The diffuser shown in Figure 1.3, simply resembles a 180° rotated C nozzle.

To achieve energy transfer in diffusers, they have an increasing section, which produces a velocity reduction of flowing gases. This velocity reduction represents a kinetic energy reduction, which is transferred to the enthalpy of the nozzle discharge fluid, thus the outflow total energy remains constant. As enthalpy is pressure energy plus internal energy, the diffuser outflow has a higher pressure and a smaller velocity than the input stream.

In any of the three cases: CD nozzle, C nozzle or diffuser, the physical principles for the design or performance assessment are the same. An interesting example of a diffuser and C nozzle combination are turbo-jet aircraft engines. In these, the cover sheath has a divergent shape air inlet and a converging shape in gases outlet. Actually, it could be said that these engines resemble a goose egg shape.

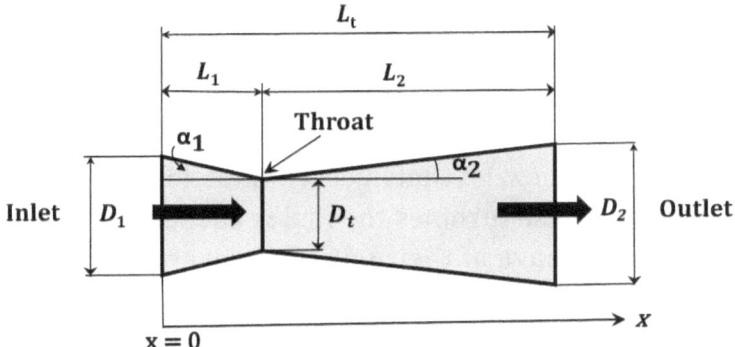

Figure 1.1. Convergent divergent nozzle

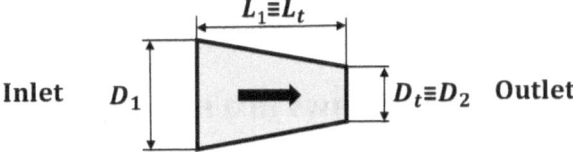

Figure 1.2. Convergent nozzle

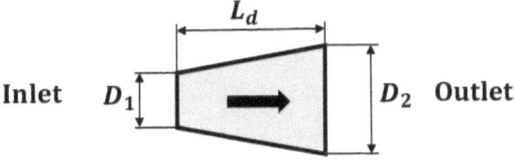

Figure 1.3. Diffuser

The incoming air flow runs at the same velocity as the aircraft plus the wind velocity that might exist at that time. The resulting velocity must be computed as a vector sum. This incoming flow is expanded into the sheath diffuser shape, which slows down its velocity and increases its pressure. This increased pressure facilitates the work of the turbine compressor, because it's added to the compressor discharge pressure. Meanwhile, the hot combustion gases, emerging from the turbine, are throated by the sheath convergent shape, thus the flow velocity increases. This increase creates the thrust that makes aircraft fly. Interestingly, both the diffuser and the outlet nozzle of turbo-jets operate with near 100% efficiency, which means that the flow friction losses in the casing, are negligible.

Gas turbines have nozzles that inject combustion gas on the first blade wheel. In these turbines the fuel is burned in combustion chambers, which have at their outlet a convergent nozzle. The hot combustion gases, driven by the GT air compressor pressure, exit through the nozzles at high velocity and impinge on the first blade wheel of the turbine. Here is where kinetic and pressure energy of gases is transferred to the turbine wheel, which converts the flow energy into the mechanical energy to drive the turbine shaft. A similar process occurs in steam turbines.

1.2. The basic physical laws in a nozzle

Note: This section is not necessary for persons who are familiar with the basic concepts of Thermodynamics.

The phenomena inside a nozzle or diffuser are based on five physical laws:

o   The principle of conservation of energy, which states that the total energy of a physical system is constant.
o   The equation of state, which mathematically links pressure, temperature and specific gravity of an ideal gas. This

equation is also applicable to combustion gases and steam, under some special conditions.

o   The principle of mass flow continuity, which states that the mass entering a control volume per unit of time, is equal to the outgoing mass flow, minus and plus the mass flow consumed or created inside the control volume. In the case of nozzles and diffusers, no retention or mass creation exists within them, so its mass flow is constant throughout the nozzle or diffuser.

o   The isentropic relations in a fluid vein.

o   The principle of conservation of the fluid's momentum, which states that the momentum is constant.

1.2.1. The principle of conservation of energy

In its simplest form, energy is a system ability to perform work. There are various forms of energy, but in the study of nozzles the main involved types of energy are the kinetic energy and the thermal energy.

In a flowing fluid the energy conservation is stated as follows.

Inflow energy = Outflow energy

Inflow energy = Potential energy (z_1) + Internal energy (u_1) +

Pressure energy $(p_1 . v_1)$ + Kinetic energy $\left(\dfrac{V_1^2}{2 \cdot g} \right)$

Outflow energy = Potential energy (z_2) + Internal energy (u_2) + Pressure energy $(p_2 . v_2)$ + Kinetic energy $\left(\dfrac{V_1^2}{2 \cdot g} \right)$ + Heat released by friction (q_f) + Net heat exchanged with the external environment (q_n) + Net mechanical work exchanged with the external environment (W_n)

The resulting expression of equalizing the inflow and the outflow energies, are conveniently simplified by the following considerations.

o Potential energies z_1 and z_2 are equal because of the nozzles short length.
o Pressure and internal energy are consolidated in only one number: the enthalpy.
o In a nozzle or diffuser no work or heat is exchanged with the external environment. Therefore, $q_n = 0$ and $W_n = 0$.

Introducing these three simplifications, the principle of energy conservation is expressed with the following equation.

Formula 1.1. Equation of conservation of energy

$$h_1 + \left(\frac{V_1^2}{2.g.J} \right) = h_2 + \left(\frac{V_2^2}{2.g.J} \right) \qquad \left| \frac{kJ}{kg} \right|$$

Where:

h_1 and h_2 are the fluid input and output enthalpies respectively

J is the mechanical equivalent of heat. Its value is: J = 101.972[(kg.m)/kJ]. In this book it was adopted:

J = 102 [(kg.m) / kJ]

g is the gravity acceleration. Its value is 9.81 $|m/s^2|$.

The output enthalpy h_2 includes the ideal enthalpy plus the friction heat released between the fluid and the nozzle inner wall. The ideal enthalpy, the friction heat and the actual enthalpy h_2, are related by following formula.

Formula 1.2. Actual discharge enthalpy

$$h_2 = h_{2\,ideal} + q_f \qquad \left|\frac{kJ}{kg}\right|$$

This enthalpy increase is associated with an entropy growth and the aftermath is that actual nozzles have lower efficiency and larger dimensions than ideal nozzles.

1.2.2. The ideal gas equation of state

The ideal gas equation of state is as follows.

Formula 1.3. Equation of state of ideal gas

$$p.v = J.R.T \qquad \left|kg.m\right|$$

Where:

R is the specific gas constant, kJ/|kg.°K |
p is the gas pressure, kg/cm² a
v is the gas specific volume, m³/kg
J is in kg.m/kJ
T is the gas temperature, °K

It is important to note that this R is not the universal gas constant, whose value is 8.315 kJ/|kmol.°K|, but the specific constant of the gas under study. The relationship between the specific gas constant R and the universal gas constant is:

Formula 1.4. Specific gas constant

$$R = \frac{R_{universal}}{MW} \qquad \left|\frac{kJ}{kg \cdot °K}\right|$$

Where MW is the molecular weight of the gas, which is expressed in |kg/kmol|.

Although steam is not an ideal gas, superheated and saturated steam transformations, can be assumed to observe the equation of state of ideal gas. This is an acceptable assumption in practice, but not always. The value of the superheated and saturated steam specific constant, is dependent on the steam pressure and temperature, thus there does not exist a unique R value for each type of steam. It is not recommended to use average values commonly seen in some technical literature because significant errors could be committed.

It should be noted that all state properties can only be determined if at least two of them are known. After the three state variables are known, the other thermodynamic properties are easily calculated, such as; enthalpy, flow velocity, kinetic energy, entropy, etc.

1.2.3. The principle of flow mass continuity

The mathematical expression of this principle, applicable to both gas and liquid, is the following.

Formula 1.5. Principle of mass continuity

$$G = A.V.\delta = \frac{A.V}{v} = \text{constant} \qquad \left|\frac{kg}{s}\right|$$

Where:

G: mass flow |kg/s|
A: area of the nozzle cross section. |m²|
V: flow velocity |m/s|
δ: fluid specific gravity |kg/m³|
V: fluid specific volume |m³/kg| = 1/δ

The A.V product represents the volume flow, which is expressed in |m³/s|. The volume flow does not comply with the continuity principle, because variations in pressure and temperature along a nozzle or diffuser change the fluid specific volume.

1.2.4. Isentropic relations

The entropy of ideal thermodynamic transformations is constant. However, in actual transformations there always exist entropy growth attributable to molecular disorder and random movement. As was said before entropy increase is caused by friction and the aftermath is a kinetic energy reduction

The heat released by friction is transferred to the nozzle body and to the fluid and thereby increases its enthalpy, as expressed in formula Formula 1.2. Actual discharge enthalpy, heat transferred to the nozzle body is negligible, hence it's neglected in this book. However, this assumption must be revised in practice, case by case. The ideal gas transformation that does not involve entropy variation is called isentropic. In this type of transformation, pressure, temperature and specific volume are mathematically related by the isentropic relations. Any process that does not dissipate heat to the environment is called adiabatic. If an adiabatic transformation is reversible it's also isentropic.

The entropy increase formula is a function of two state variables. The most commonly used are the initial and final temperatures and pressures. In this case, the applicable formula is as shown below.

Formula 1.6. Entropy increase

$$S_2 - S_1 = c_p \cdot \ln\frac{T_2}{T_1} - R.\ln\frac{p_2}{p_1} \qquad \left|\frac{kJ}{kg.°K}\right|$$

The specific heats c_p and c_v and the specific gas constant R, are related by following formula.

Formula 1.7. Gas constant and specific heats

$$R = c_p - c_v = c_p \frac{k-1}{k} \qquad \left| \frac{kJ}{kg.°K} \right|$$

Where k is the specific heats ratio, therefore, by replacing

Formula1.7in Formula 1.6, the isentropic condition is obtained:

Formula 1.8. Isentropic condition

$$\ln \frac{T_2}{T_1} = \frac{k-1}{k}.\ln \frac{p_2}{p_1} \qquad \left| \frac{kJ}{kg.°K} \right|$$

Clearing the T_1/T_2 ratio from Formula 1.8, the following isentropic relations are obtained.

Formula 1.9. Isentropic relations

$$\frac{T_1}{T_2} = \left(\frac{p_1}{p_2} \right)^{1-\frac{1}{k}} = \left(\frac{v_2}{v_1} \right)^{k-1}$$

These relations are intensively used in many thermodynamic calculations and, of course, in the nozzle engineering as well. The importance of the isentropic relations is that they allow the thermodynamic properties determination at any nozzle section, provided that the same properties are known at any other location inside the nozzle, usually at the inlet for example.

In general, p_1 and T_1 are known, hence, the specific volume at the inlet is given by the equation of state. Once the three state variables are known, the application of the isentropic relations allow calculating their equivalent property at any other section of the nozzle. Generally, the isentropic relation is used to calculate the state properties at the throat and at the outlet, on the basis of their values at the inlet. These relationships are also applicable to superheated and saturated steam, although the latter at a much smaller extent.

1.2.5. Enthalpy and internal energy formulas

Enthalpy is thermal energy formed by two components: internal energy and pressure energy. In an ideal gas, enthalpy is expressed only as a function of temperature. The applicable formulas are as follows.

Formula 1.10. Enthalpy. General formula

$$h = u + \frac{p.v}{J} \qquad \left|\frac{kJ}{kg}\right|$$

Formula 1.11. Enthalpy of ideal gas

$$h = c_p.T \qquad \left|\frac{kJ}{kg}\right|$$

Formula 1.12. Internal energy

$$u = c_v.T \qquad \left|\frac{kJ}{kg}\right|$$

Where:

u: internal energy, kj/|kg.°K|
c_p: specific heat at constant pressure, kj/|kg.°K|
c_v: specific heat at constant volume, kj/|kg.°K|

Formulas 1.3,1.6, 1.9, 1.10, 1.11 and 1.12 demonstrate that specific volume, entropy, enthalpy and internal energy are easily calculated by measuring pressure and temperature.

1.3. Energy conversion in nozzle flow

Nozzle's purpose is the conversion of the available thermal energy in the fluid into kinetic energy of the outgoing jet. The total available energy is defined as the difference between the input and the output enthalpies of the fluid. Then the kinetic energy is increased at the expense of the total available energy in the fluid. However, part of the total available energy is not converted into kinetic energy due to friction losses between the fluid and the nozzle inner walls. A complete definition of total available energy is in sections 2.1 and 2.5.

Accompanying this energy exchange there exists, of course, a strong expansion of the fluid to increase its velocity. This phenomenon is manifested by the increase of specific volume or loss of specific gravity. While this expansion occurs, fluid temperature comes down. This temperature reduction is opposed by the friction between the fluid and the nozzle inner walls, which add heat to the fluid. Then this additional heat reduces the temperature drop within the nozzle. The amount of returned heat to the fluid is typically in the range of 4% to 19% of the total available energy. These figures are not insignificant to the overall turbine efficiency and must be taken into account in the nozzle design or operation calculations.

Performance curves have been calculated in file 1.1. Figures 1.4 to 1.8. Performance curves.xlsx. This file is a complete calculation model of a CD nozzle. All calculation procedures are in blue color tabs and figures are in yellow color tabs. It is not needed at this time to read the calculation procedures, because these will be explained in Chapters 5, 6 and 7.

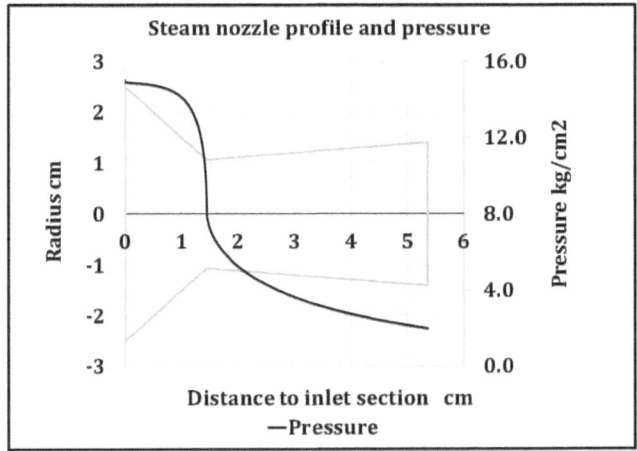

Figure 1.4. Nozzle profile and pressure distribution

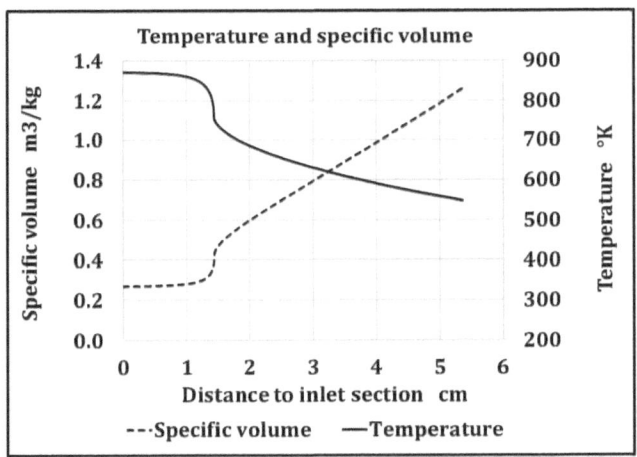

Figure 1.5. Steam temperature and specific volume distribution

Figure 1.4 and Figure1.5represent the distribution of thermodynamic state properties in a CD nozzle, versus its distance to inlet. This representation assumes, without much error, that the nozzle flow is one-dimensional. It must be noted that all properties experience a sudden variation at the throat, where the flow is constricted.

The same is done in Figure 1.6 with the kinetic energy and the fluid enthalpy. Note that the enthalpy and kinetic energy curves are complementary so that their sum remains constant with respect the principle of energy conservation. This plot clearly shows the energy interchange between enthalpy and kinetic energy. The last grows at the expense of the fluid enthalpy.

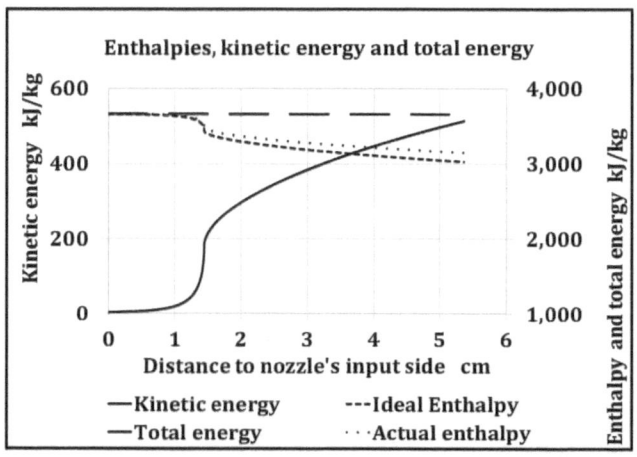

Figure 1.6. Steam enthalpies and energies

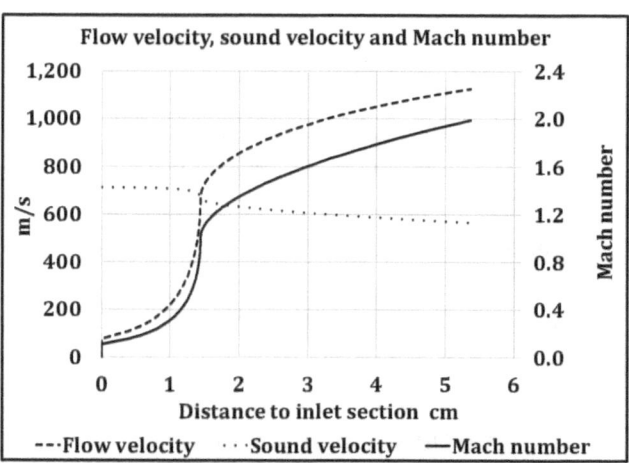

Figure 1.7. Steam and sound velocities
and Mach number distribution

Figure1.7 represents the flow velocities and their Mach number ($M_x = V_x/V_{sound\ at\ x}$). This figure indicates that the outgoing velocity is twice the sound velocity.

1.4. Flow velocity behavior in nozzles

Inlet velocity in C nozzles is usually negligible in practice, but immediately after the inlet section the fluid strongly accelerates. The associated discharge velocity can almost match the speed of sound, but cannot overcome it. Instead, in CD nozzles the inlet velocity is also negligible, but in the divergent part the flow exceeds the speed of sound. (See Figure 1.7. This figure shows a case where ideal velocity reaches 1,127 m/s).

In the case of saturated steam, the flow velocity at the throat is about 450 m/s. A typical velocity value in gases is similar to steam or even higher. This value varies very little with the inlet pressure of the nozzle. In this example the ideal throat velocity is 668 m/s.

Fluid velocity at the throat abruptly increases while pressure also abruptly falls. See Figure 1.4 and Figure1.7. In the constricted section the Mach number is equal to 1 so the circulation time throughout the nozzle is very small; on the order of few milliseconds.

A strong expansion (specific volume growth) of the fluid takes place in the divergent part, at a higher rate than the fluid's velocity. In this part enthalpy is converted into kinetic energy to a higher extent than it does in the convergent part. The result is that the outgoing velocity may double in value despite the cross section increase.

1.5. Flow critical properties

When the fluid velocity matches the speed of sound, it is said that the nozzle is operating in critical conditions. In this case, pressure, temperature, specific volume and velocity at the throat, are called critical thermodynamic properties. To design or evaluate a nozzle, a significant property that needs to be known is the pressure distribution inside the nozzle. In the design stage, the comparison between the discharge pressure versus the critical pressureanticipates the type of nozzle that is needed and also the discharge stream shape, which may be a uniform or a scattered flow.

1.6. Discharge jet shape

It was established in previous sections that the nozzle discharge jet strikes the turbine blades and transfers its energy to the turbine rotor. But for this to occur the discharge jet shape must have a uniform tube shape and not a plume shape going out of the nozzle. The shape of the flow depends on the velocity and turbulence within the nozzle and in the discharge stream. A subsonic or sonic flow has a uniform tube shape, unlike supersonic flow shape, which is dispersed.

To prevent dispersion of a supersonic flow, it is confined in the divergent part of a CD nozzle. Therefore, the divergent part encloses the outgoing flow and minimizes its dispersion. This type of nozzle was invented by the Swedish Engineer De Laval (1845-1913) in 1890, to obtain uniform flows at supersonic velocities in the nozzle discharge.

1.7. Specific mass flow

For this section refer to file 1.1. Figures 1.4 to 1.8. Performance curves.xlsx.

Another parameter that characterizes a nozzle operation is its mass flow density distribution or specific mass flow. The mass flow density is the ratio of mass flow over the cross section area where flow is passing by. Figure 1.8 shows the flow along with the sections area curve. It can also be defined as the amount of mass flowing through the nozzle per unit of time and per unit of area of the cross section. Its mathematical expression is:

Formula 1.13. Specific mass flow

$$g_{sx} = \frac{G}{A_x} \qquad \left| \frac{kg}{s \cdot m^2} \right|$$

Where g_{sx} is the mass flow density, or specific mass flow, and A_x is the nozzle cross section area located at a distance x from the inlet. By the principle of continuity, G is constant at any location x, but it is not the same with the specific mass flow, because of the nozzle changing area. Specific mass flow g_{sx} distribution is plotted versus the abscissa x in Figure 1.8. This graph shows that the curve presents a sharp discontinuity at the throat. This discontinuity point is also the maximum value of g_{sx}.

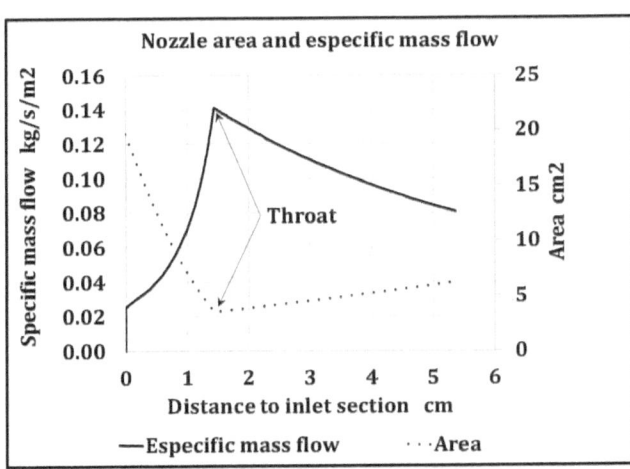

Figure 1.8. Specific mass flow and sections area curves

1.8. Turbulence in nozzles

Within a flow, fluid particles are moved by different type of forces. However, two of them define whether the flow is formed by continuous and quiet trajectories or by chaotic movements. These forces are:

○   Viscous forces
○   Inertial forces

Viscous forces are produced by friction between fluid layers and quiet flow is the aftermath of predominant viscous forces. This regime is named laminar. A good example of high viscosity flow is a honey poured from a jar to another container.

Inertial forces are described by the 2nd law of Newton. They are influenced by particles mass and accelerations. At high velocities fluid particles experience velocity fluctuations, which finally produces a chaotic flow. This is the case where inertial forces are predominant.

In a laminar regime, fluid particles are moved ahead by forces produced by pressure difference between the nozzle's input and output. As flow is expanding (See specific volume curve in Figure 1.5) particles also move transversally towards nozzle's wall. However, low viscosity fluids and high linear velocities create a chaotic movement formed by vortices and eddies, mainly next to the inner walls of the nozzle. This chaotic regime is superimposed to the laminar regime. The result is a turbulent regime, which is a very complex phenomenon, difficult to model and still there does not exist a general solution in Fluid Mechanic's theory.

Turbulence is very important to nozzles because it creates a friction between the flow and the nozzle inner walls that reduces flow velocity. This is a similar effect to the friction created by the inner wall roughness. Actually, both effects; turbulence and friction, add one to the other and their compound impact on the flow is represented by only one number, known as friction factor. The impact of friction and turbulence is the reduction of the kinetic energy of the outgoing flow. This impact is minimized by means of a proper nozzle design. This means that the nozzle profile must accommodate the flow with minimum turbulence. Actually, this means that the flow shape fits, as much as possible, the nozzle profile.

To determine whether a flow is turbulent or not, Fluid Mechanics states that the Reynolds number must be calculated and compared with a predetermined value, above which flow regimes are turbulent. In pipes, it is assumed that flow is turbulent for values of Reynolds number higher than 4,000.

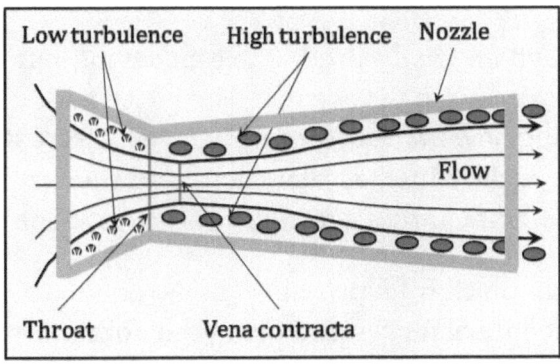

Figure 1.9. Turbulent flow inside a CD nozzle

Experience shows that rapid converging passages have very little turbulence. That's why the converging part of actual nozzles is very short. Instead, divergent part of the nozzle is gradually enlarged to avoid high turbulence. It's important to note that turbulence in nozzles can be mitigated but not canceled. SeeFormula 1.9.

In the nozzle design and assessment explained in this book, the Reynolds number formula for pipes is used. This criterion assumes that the nozzle is composed of many pipes of very small length each, where the Reynolds number is calculated one pipe at a time. The result will be a set of Reynolds numbers, which represent the distribution along the nozzle or the turbulence severity point by point.

The result is indicative of a flow behavior within the nozzle, which usually is turbulent or complete turbulent. See the explanation of the turbulence classification in section 4.1.1. Eventually, this procedure is used to calculate the compound friction factor throughout the nozzle or to compare different flow regimes to each other, to define which nozzle has a higher or smaller turbulence.

1.9. Shock waves in CD nozzles

C nozzles have an easily predictable behavior, but it does not happen the same with CD nozzles. The latter have a supersonic and expanding flow in the diverging part, which creates complex phenomena, such as high accelerations and shock waves. These phenomena significantly affect nozzle efficiency. See section 2.2.

If the actual discharge pressure p_2 is equal to the design pressure, the internal pressure along the nozzle throat sharply drops at the throat and then smoothly decays in the divergent part, similar to an exponential curve, until it reaches the outlet section. All other thermodynamic properties follow the pressure curve according to applicable physical laws. Furthermore, all other properties can be expressed as a function of pressure and this makes possible to create for any property a table or curve versus the abscissa x or pressure p_2.

But in practice, it often happens that discharge pressure is regulated to values other than design specifications. If operational reasons oblige to operate with outlet pressure higher than a specific value, a conflict between thermodynamic state properties is created. These phenomena is discussed in Chapter 4. Nature works out the conflict by creating, inside the nozzle, a properties discontinuity lamina, which is perpendicular to the flow direction, called normal shock wave. Also, if the design pressure curve is below the receiving environment pressure, the shock wave finally matches, outside the nozzle, the discharge pressure to that of the receiving environment. In this case oblique shock waves are formed within the outgoing flow.

Importantly, it's not that an internal shock wave may be formed but that inexorably a shock wave will be formed whenever the discharge pressure is higher than a specific limit pressure. From there comes the importance of operating nozzles as close as possible to the design conditions.

Discontinuity formed by a normal shock wave is manifested as a sudden increase in pressure and temperature and a sudden velocity drop, at the shock wave exit. The velocity regime after the shock wave is no longer supersonic but subsonic. That's why the divergent part of the nozzle works as a diffuser after a shock wave, which means that pressure and temperature increase.

Another important aspect is that the thermodynamic transformation that occurs between both sides of the shock wave is irreversible. Therefore, the entropy at the exit face of the wave is higher than at the input.

For a turbine operation, a shock wave is not a minor issue, because it reduces the kinetic energy of the discharge flow and may even prevent the flow to circulate. The reason for this is that a shock wave works as if the outside pressure had formed a plug inside the nozzle, this plug is finally overcome, but at the expense of an entropy increase of the flow and a sharp change in the fluid thermodynamic properties. That plug is generated and destroyed alternatively, creating a zone of instability.

The thickness of the discontinuity lamina or shock wave is about the average spatial distance between molecules. For example, the air thickness is about 2×10^{-5} to 5×10^{-5} mm, depending on the temperature. Anyway, for the nozzle study the shock wave thickness is considered nil, for it is just a limit after which the fluid thermodynamic properties suddenly change. Between the nozzle inlet and the shock wave inlet all the thermodynamic variables exactly match the design conditions. No changes happen to properties distribution pattern in that space.

The shock wave location within the nozzle is as close to the throat the higher is the external pressure. That's why the shock wave travels to the nozzle outlet whenever the external pressure is going down. Oblique shock waves formed after the nozzle outlet produce sudden changes in pressure, temperature and flow velocity outside the nozzle. These shock waves equal the discharge jet pressure with the external pressure, just as normal shock waves do it inside the nozzle.

1.10. Flow choking

Fluid Mechanics states that the flow velocity between two points of a fluid vein depends on the pressure difference between them. The higher this difference, the higher the fluid velocity. The application of this behavior to nozzles suggests that the velocity will steadily increase if the input pressure remains constant and discharge pressure p_2 steadily decreases, but this is correct up to a point. There is a limit to the flow velocity growth, which happens when the flow approaches or reaches the speed of sound. From this moment, even though the nozzle discharge pressure is decreased, the speed at the throat remains constant and equal to that of the speed of sound. This phenomenon of stalling speed at the throat is known as strangled or choked flow.

According to Osborne Reynolds's ingenious explanation, the reason for the choking phenomenon is the elastic behavior of gas molecules. This elasticity facilitates waves propagation produced by perturbations of the upstream pressure, moving at the speed of sound. And as the fluid also moves at that speed in opposite direction, both velocities are canceled, preventing changes in the output pressure p_2 to affect the throat pressure or the discharge pressure in C nozzles. Because of these equal velocities in opposition, flow at the throat has no way to sense that there has been a change in the discharge pressure; consequently, its speed remains constant and equal to speed of sound. This phenomenon is very well known in the oil and gas fields, where it is necessary to maintain constant pressure at the wellhead, so that oil or gas extraction is not disturbed by surface facilities. That is why bottlenecks are placed in extraction pipes discharge. These bottlenecks are usually known as chokes and efficiently isolate the extraction operation from the surface facilities.

When the throat flow reaches the speed of sound, pressure at the throat is known as critical pressure and the flow velocity is known as critical velocity.

The critical pressure is very simple to calculate for each type of fluid. Its value is only a function of the fluid's specific heats ratio ($k = c_p/c_v$), just as are all the other thermodynamical properties. The choked flow formulas of Chapter 3, demonstrate that the critical pressure is between 48% and 58% of the inlet pressure. Its exact value depends on the type of fluid flowing through the nozzle. See the column of r_{pc} values inTable 1.1.

Critical condition exists whenever the actual throat pressure p_t is equal to the critical pressure p_c. When this is accomplished, the nozzle is necessarily a CD type nozzle, and the flow velocity equals or exceeds that of sound in the divergent part. Instead, in C type nozzles, speed of sound is never reached, which means that C nozzles never operate in critical conditions. In C nozzles p_2 is always greater than p_c, whereas in CD nozzles p_2 is always smaller than or equal to p_t and therefore, always operate in critical regime. If a CD nozzle is not operating in critical conditions, its use is meaningless and should be replaced by a C nozzle.

In CD nozzles, condition of choked flow is expressed as follows:

Formula 1.14. Pressure condition of choked flow for a CD nozzle

$$p_2 \leq p_t = p_c \qquad \left| \frac{kg}{cm^2} \right|$$

The critical state variables (p_c, v_c, T_c) at the throat, are directly proportional to the same variables at the nozzle inlet. The proportionality constants are called critical ratios of pressure, specific volume and temperature respectively. The critical ratios formulas are:

Formula 1.15. Critical ratio formulas

$$r_{pc} = \frac{p_c}{p_1} \qquad r_{tc} = \frac{T_c}{T_1} \qquad r_{vc} = \frac{v_c}{v_1}$$

These ratios are constants and do not depend on the nozzle geometry but on the fluid specific heat ratio k. According to Formula 1.14 and 1-15 for critical flow conditions to exist the nozzle pressure ratio should be lower than its critical pressure ratio. This condition is derived by dividing all terms of Formula 1.14 by p_1.

Formula 1.16. Choked flow condition defined
by pressure ratios for a CD nozzle

$$r_{pn} \leq r_{pt} = r_{pc}$$

Where r_{pn} is the nozzle pressure ratio p_2/p_1

As said before a C type nozzle is never choked for the discharge velocity never reaches the speed of sound. That means that in a C nozzle it is always: $p_2/p_1 > r_{pc}$.

The values of pressure, temperature and specific volume critical ratios, for different working substances in gas and steam turbines, are shown in Table 1.1. The critical ratio formulas are demonstrated in Chapter 3.

1.11. Example of a critical ratio calculation

For this section refer to file 1.3. Figure 1.10. Velocity and mass flow versus p1.xlsx.

Let it be a CD nozzle conveying combustion gas. The inlet pressure p_1 is equal to 15 kg/cm² a. The critical pressure is obtained by reading the value of r_{pc} in Table 1.1 and multiplying it by the value of the inlet pressure. In this case, the critical pressure at the throat is:

p_c = 0.5404 x 15 = 8.11 kg / cm2.

For any discharge pressure smaller than 8.11 kg/cm2 a, the flow is choked and therefore, its throat speed is constant and equal to the speed of sound. As in this case the discharge pressure is 2 kg/cm2 a, which is smaller than the critical pressure, the nozzle type is CD.

Figure 1.10 shows the flow velocity and mass flow curves versus pressure at the throat. This graph has been calculated with formulas to be explained in Chapter 3. These curves are similar because they are related by Formula 1.5. Their plot demonstrate that flow velocity and mass flow have a non-linear relationship for discharge pressures higher than critical pressure (or $r_{pt} \geq r_{pc}$) because the specific volume varies with pressure changes. Also, Figure 1.10 demonstrates that in this region ($p_2 \geq p_c$) the speed grows at a smaller rate than the mass flow.

There is a way to increase or decrease the mass flow under choked flow conditions, which is done by adjusting the nozzle inlet pressure. If the inlet pressure changes, the fluid gravity also changes and hence, its mass flow. In that case the mass flow curve moves vertically, up or down, depending on the sign of the applied pressure change to p_1. Instead, by adjusting p_2 when the flow is choked, it is not possible to change the mass flow.

Table 1.1. Critical ratios table for different gases and steam types

Nozzle critical ratios					
Fluid	Steam title	k = cp/cv	rpc = pc/p1	rtc = Tc/T1	rvc = vc/v1
Wet steam	0.800	1.115	0.5815	0.9456	1.6260
Wet steam	0.850	1.120	0.5805	0.9434	1.6251
Wet steam	0.900	1.125	0.5795	0.9412	1.6242
Wet steam	0.950	1.130	0.5785	0.9390	1.6232
Saturated steam	1.000	1.135	0.5774	0.9368	1.6223
Superheated steam		1.300	0.5457	0.8696	1.5934
Combustion gases		1.330	0.5404	0.8584	1.5885
Air		1.400	0.5283	0.8333	1.5774
Monoatomic gases		1.667	0.4871	0.7499	1.5396

See file 1.4. Table 1.1. Critical ratios.xlsx

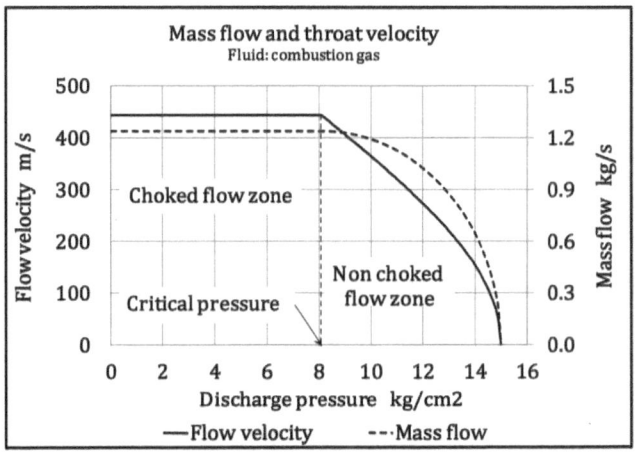

Figure 1.10. Velocity and mass flow behavior
versus the discharge pressure

CHAPTER 2

AVAILABLE ENERGY AND EFFICIENCY OF NOZZLES

2.1. Available energy for $V_1 = 0$

The available energy is the maximum amount of energy that a substance can convert from one form to another. Turbine design seeks to maximize the flow energy transfer from the fluid to the drive wheels. In nozzles, the goal is to maximize the conversion of total available energy into kinetic energy. The maximum conversion efficiency is achieved when the transformation is reversible, meaning that no friction exists within the nozzle. It is an ideal fact, unattainable in practice because friction is absolutely an unavoidable phenomenon. The irreversibility of a fluid transformation is measured by the entropy increase.

Formula 2.1demonstrates that the fluid enthalpy change is equal to its kinetic energy variation. This variation is the ideal total available energy E_a.

Formula 2.1. Ideal total available energy or ideal TAE

$$E_{a\,ideal} = h_1 - h_{2ideal} = \left(\frac{V_{2ideal}^2}{2 \cdot g \cdot J} \right) - \left(\frac{V_1^2}{2 \cdot g \cdot J} \right) \left| \frac{kJ}{kg} \right|$$

The idealsubscriptdenotes the absence of friction.

As reality demonstrates that the initial velocity V_1 is very small compared to V_2, V_1 is usually negligible. The actual speed V_2 is smaller than the ideal value because of the existing friction within the nozzle. Therefore, in the actual available energy formula; enthalpy h_{2ideal} has to be replaced by the actual enthalpy h_2. As was said before, friction makes enthalpy h_2 to be higher than h_{2ideal} and Consequently, the actual total available energy E_a is lower than $E_{a\ ideal}$.

Formula 2.2. Actual total available energy or TAE

$$E_a = \left(\frac{V_2^2}{2 \cdot g \cdot J} \right) = h_1 - h_2 \qquad \left| \frac{kJ}{kg} \right|$$

Given that initial velocity is zero, the available energy equals the outlet kinetic energy. If initial velocity is not zero, the inlet kinetic energy must be deducted from the available energy.

As a gas enthalpy equals $c_p.T$, the available energy for gas nozzles is written as follows.

Formula 2.3. Available energy in a gas nozzle

$$E_a = c_p \cdot \left(T_1 - T_2 \right) \qquad \left| \frac{kJ}{kg} \right|$$

Formula 2.3is not applicable to steam. The correct formulas for steam are the following:

Formula 2.4. Superheated steam enthalpy. Mollier formula

$$h_b = 3.186 \times \left[0.47 \times \left(T - 273 \right) - \frac{201.96}{\left(\frac{T}{100} \right)^{\frac{10}{3}}} \times p - \frac{1.248 \times 10^{12}}{\left(\frac{T}{100} \right)^{14}} \times \left(\frac{p}{100} \right)^3 + 595 \right] \qquad \left| \frac{kJ}{kg} \right|$$

Formula 2.5. Saturated steam enthalpy.

$$h_s = 419.728 \cdot p^{0.257071} - 5.2360 \times 10^{-3} \cdot p^3 + 0.6506 \cdot p^2 - 29.2456 \cdot p + 2,255.95 \qquad \left|\frac{kJ}{kg}\right|$$

Formula 2.6. Wet steam enthalpy.

$$h_h = 419.728 \cdot p^{0.257071} + x \cdot \left(-5.2360 \times 10^{-3} \cdot p^3 + 0.6506 \cdot p^2 - 29.2456 \cdot p + 2,255.95\right) \qquad \left|\frac{kJ}{kg}\right|$$

Where in Formula 2.6 letter x is the steam title (or steam quality).

Any of these three formulas is used to calculate steam enthalpies h_1 and h_2. Their accuracy is good in the range of 1-50 kg/cm2 a.

The outlet kinetic energy over the total incoming energy h_1, expressed in %, is called the conversion efficiency. This efficiency is calculated with the following formula. See demonstration in section 8.2.

Formula 2.7. Energy conversion efficiency

$$e_c = \frac{E_{k2}}{h_1} = 1 - \left(\frac{p_2}{p_1}\right)^{1-\frac{1}{k}}$$

Where e_c is the nozzle conversion efficiency, E_{k2} is the kinetic energy at the outlet $(V_2^2/(2.g.J))$ and p_1 and p_2 are the inlet and discharge pressures respectively. The e_c graph versus the compression ratio p_2/p_1 is shown in Figure 2.1. This graph represents e_c for three different fluids; that is ideal gas, combustion gas and superheated steam. Steam curve is not exact because steam is not an ideal gas, hence, the representation of Figure 2.1 must be considered just an acceptable approximation. The superheated steam curve demonstrates that input enthalpy conversion is lower than ideal gas or combustion gas. This lower conversion efficiency is due to the presence of evaporated water in the steam mass.

For this graph refer to file2.1. Figure 2.1 Conversion efficiency of h1.xlsx.

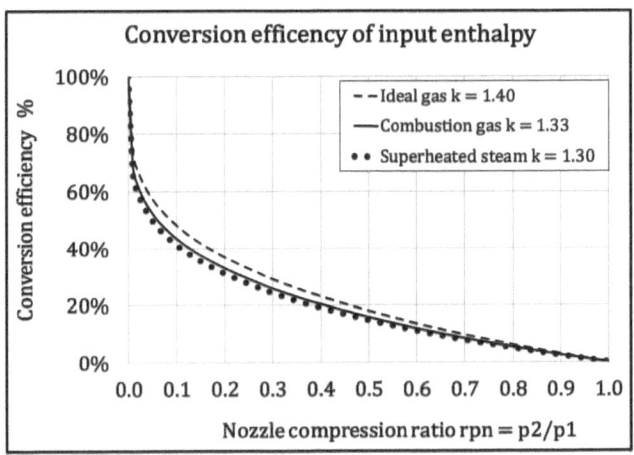

Figure 2.1. Conversion efficiency of inlet enthalpy h_1. Steam curve is an approximation

From Figure 2.1 it is concluded that conversion efficiency has a sudden drop for values lower than about 5% of r_{pn}. After this initial sudden fall, efficiency slowly drops for all substances until it becomes zero when the discharge pressure equals that of the input. Figure 2.1 also shows that conversion efficiency is higher the lower is compression ratio.

2.2. Energy efficiency of a nozzle

Energy efficiency of a nozzle is equal to the actual kinetic energy of the outflow over the ideal kinetic energy in that location, or equal to the total actual enthalpy change over the total ideal enthalpy change.

Formula 2.8. Nozzle efficiency

$$e_n = \frac{E_{k2}}{E_{k2ideal}} = \left(\frac{V_2}{V_{2ideal}}\right)^2 = \frac{h_1 - h_2}{h_1 - h_{2ideal}}$$

The relationship between V_2 and $V_{2\,ideal}$ of Formula 2.8 is smaller than 1, and is known as velocity coefficient. It is designated by φ and its relationship with energy arises from Formula 2.9.

Formula 2.9. Velocity coefficient

$$\varphi = \frac{V_2}{V_{2ideal}} = \sqrt[2]{e_n}$$

Formula 2.9 shows that nozzle efficiency e_n equals the square of the velocity ratio φ. The procedures and formulas to calculate the velocity coefficient are presented inChapter 4.

2.3. Friction losses and enthalpy discharge

As was said before, the friction heat losses released inside the nozzle is transferred to the fluid. This transference increases the discharge enthalpyfrom h_{2ideal} to h_2, which is the actual enthalpy. As for practical purposes it is assumed that no part of this heat is transferred to the nozzle body. Hence, the discharge enthalpy increment equals the total friction released inside the nozzle. This total heat loss incorporated to the fluid is designated q_f, which is equal to the $(h_2 - h_{2\,ideal})$ difference.

From Formula 2.8 it is possible to express q_f as a function of the nozzle efficiency e_n.

Formula 2.10. Heat loss or reheat

$$q_f = h_2 - h_{2ideal} = E_{k2ideal} - E_{k2} = (1-e_n)\cdot(h_1 - h_{2ideal})\left|\frac{kJ}{kg}\right|$$

Actual discharge enthalpy h_2 can also be cleared from Formula 2.10 and the result is the following expression.

Formula 2.11. Actual discharge enthalpy

$$h_2 = h_1 - e_n \cdot E_{a\,ideal} \qquad \left|\frac{kJ}{kg}\right|$$

These last two formulas are useful to calculate the nozzle heat loss (reheat) and their actual discharge enthalpy.

2.4. Example of efficiency calculation

Determine:

o   Total available energy
o   Conversion efficiency
o   Nozzle efficiency
o   Actual discharge velocity
o   Friction losses

The nozzle operating conditions are:

o   Inlet temperature: 1300°K
o   Actual outlet temperature: 800°K
o   Specific heat coefficient c_p: 1.1 kJ/|kg.°K|
o   Input velocity V_1: 0
o   Ideal discharge velocity: 1,100 m/s

Calculations:

○ Actual enthalpies:

$h_1 = 1.1 \times 1,300 = 1,430 \; |kJ/kg|$

$h_2 = 1.1 \times 800 = 880 \; |kJ/kg|$

○ Energy conversion:

$E_d = E_{k2} = 1,430 - 880 = 550 \; |kJ/kg|$

$e_c = 550/1,430 \times 100 = 38.46\%$

○ Actual discharge velocity:

$$V_2 = \sqrt[2]{2 \cdot g \cdot J \cdot (h_1 - h_2)} = \sqrt[2]{2 \times 9.81 \times 102 \times 550} = 1,049 \; |m/s|$$

○ Velocity coefficient:

$\varphi = 1,049/1,100 = 0.9536$

○ Nozzle efficiency:

$e_n = 0.9536^2 \times 100 = 90.94\%$

○ Ideal enthalpy at the outlet is cleared from Formula 2.11

$h_{2 \, ideal} = 1,430 - 550/0.9094 = 825 \; |kJ/kg|$

○ Friction losses:

$q_f = 880 - 825 = 55 \; |kJ/kg|$

2.5. Total available energy and discharge enthalpy for $V_1 \neq 0$

If the input velocity cannot be neglected, the consequent input kinetic energy must be added to total available energy. After this replacement the total available energy formula is the following.

Formula 2.12. Total available energy for $V_1 \neq 0$

$$E_a = h_1 - h_2 + \frac{V_1^2}{2 \cdot g \cdot J} \qquad \left|\frac{m}{s}\right|$$

Similarly, the actual discharge enthalpy is given by the following formula, derived from Formula 2.11 and 2.12.

Formula 2.13. Actual discharge enthalpy

$$h_2 = h_1 - e_n \cdot \left(h_1 - h_{2ideal} + \frac{V_1^2}{2 \cdot g \cdot J} \right) \qquad \left|\frac{kJ}{kg}\right|$$

However, Formulas 2.12 and 2.13 are seldom used in practice because of the negligible values of input velocity V_1 compared to flow velocities inside the nozzle. It doesn't happen the same in gas turbines, because the air compressor discharge flow can reach up to 150 m/s. Nonetheless the combustion chamber might mitigate this issue.

2.6. Steam expansion in a nozzle

The steam and gas turbine nozzles share common formulas and calculation procedures. However, there are some differences that should be considered when evaluating or projecting a steam nozzle.

The fluid expansion in a nozzle, represented by a significant increase of its specific volume, is the mother mechanism of a nozzle operation. The steam expansion is calculated with formulas and also with the Mollier diagram. In this abacus, coordinates are the steam enthalpy and entropy. A group of curves represent lines of constant values of different thermodynamic properties, such as pressure, temperature, specific volume and steam title (or steam quality). The curves intersection at any point completely define the thermodynamic state of the steam at that point. Therefore, in this grid of constant properties curves, paths followed by steam transformations can be traced. The fundaments and use of Mollier diagram is very well known so no theory or usage procedure are presented in this book. However, it will be used here to explain graphically the steam expansion in a nozzle.

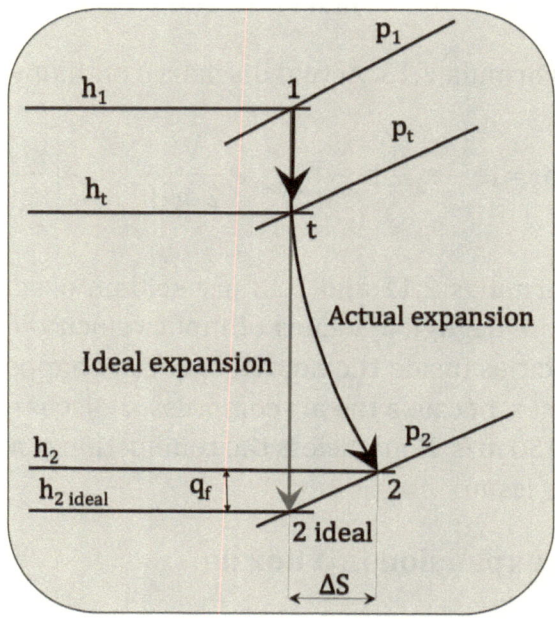

Figure 2.2. Nozzle expansion in the Mollier diagram

In Figure 2.2 the h_x horizontal lines represent constant enthalpy. The slanted lines p_x represent constant absolute pressure. The vertical line 1-t represents the expansion in the convergent nozzle part that is between the inlet and the throat. Due to the short length of this part, friction losses are negligible, therefore, the ideal and actual expansion lines coincide in this especial case. The vertical line t-2$_{ideal}$ represents the ideal expansion in the divergent part. There is no friction in this transformation.

The curve t-2 is the actual expansion in the divergent part. Because of friction there exists an entropy increase ΔS. As was seen in section 2.3, the q_f heat released by friction incorporates to the steam, which increases the discharge enthalpy. The curve t-2 is called condition curve and is aimed at showing the successive thermodynamic states of the working substance.

The heat loss q_f is read directly from the vertical distance between point 2 and point 2_{ideal}. Point 2 represents the blades location, where steam is discharged from the turbine. This may be the condenser or any other device where remaining steam energy is harnessed.

CHAPTER 3

VELOCITY AND MASS FLOW IN NOZZLES

To drive a turbine rotor, the kinetic energy of a nozzle outflow is transferred to the rotor, hence, this last energy should have a proper value according to the expected turbine power. Therefore, to calculate the kinetic energy of the outgoing flow, either in the design phase or during the operation, the discharge flow velocity is required. This is calculated with Formula 3.1, which states that the discharge flow velocity depends on the actual enthalpy change inside the nozzle, or with the Saint Venant - Wantzel equation (Formula 3.4), later derived also by Weisbach. The Saint Venant - Wantzel equation is only valid for a non-strangled gas flow, but Formula 3.1 is applicable to choked or non-choked flow.

For this Chapter, refer to file 3.1. Figures 3.1 and 3.2. Velocity and mass flow versus pressure ratio.xlsx.

3.1. Flow velocity calculation with enthalpy drop

Assuming that friction and input velocity are negligible, the output flow velocity is derived from the equation 1.1 of the energy conservation principle. The result is following formula.

Formula 3.1. Ideal discharge velocity

$$V_{2\,ideal} = \sqrt[2]{2 \cdot g \cdot J \cdot \left(h_1 - h_{2\,ideal}\right)} \qquad \left|\frac{m}{s}\right|$$

This formula is applicable to any type of flow, whether or not strangled and also valid for ideal gases and steam.

Gas enthalpy is calculated with Formula 1.11. The value of specific heat c_p is temperature dependent, but in this case this property was not considered, which represents an acceptable approximation for the nozzle study. However, where high accuracy calculation is required, it is not acceptable to disregard the c_p dependency from temperature. This is the case in GT cycles calculation.

Steam enthalpy values are calculated with Formulas 2.4, 2.5 and 2.6.

To account for the friction impact on flow velocity, Formula 3.1 has to be multiplied by the velocity coefficient φ. Then the actual velocity formula is the following expression.

Formula 3.2. Actual discharge velocity

$$V_2 = \varphi \cdot \sqrt[2]{2 \cdot g \cdot J \cdot \left(h_1 - h_{2\,ideal} \right)} \qquad \left| \frac{m}{s} \right|$$

However, an equally valid form of this formula is obtained by replacing $h_{2\,ideal}$ with the actual h_2, which is higher than the ideal enthalpy value. See formulas 2.7 and 2.8. The result is following formula.

Formula 3.3. Actual discharge velocity

$$V_2 = \sqrt[2]{2 \cdot g \cdot J \cdot \left(h_1 - h_2 \right)} \qquad \left| \frac{m}{s} \right|$$

Other forms of Formula 3.3 are presented in this chapter.

3.2. Saint Venant - Wantzel equation for ideal gases

Between two containers at different pressure, separated by a thin wall having an orifice, a flow is established through this orifice. As is known, the fluid goes from the higher pressure vessel to the lower pressure vessel. Fluid Mechanics states that the minimum flow section area is not at the orifice but shortly after leaving the orifice (vena contracta) and immediately expands again. (See Figure 1.9). In the nineteenth century, Jean Claude Barre de Saint-Venant and Pierre Wantzel, derived a formula to calculate the velocity of an orifice outgoing flow, which is applicable to gas and steam nozzles. This equation is today known as the Saint Venant - Wantzel equation. Its general expression for an ideal gas is derived in section 8.3.

Formula 3.4. Saint Venant–Wantzel equation

$$V_x\left(r_{po}\right) = \sqrt[2]{\frac{2 \cdot g \cdot k}{k-1}\left[1-\left(r_{po}\right)^{\frac{k-1}{k}}\right]} \times \sqrt[2]{J \cdot R \cdot T_1} \qquad \left|\frac{m}{s}\right|$$

Subscript "o" stands for orifice. A graphic interpretation of orifice in CD and C nozzles is in Figure 3.1.

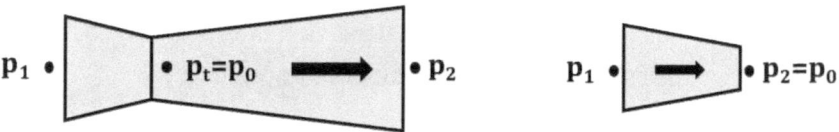

Figure 3.1. Saint Venant - Wantzel equation.
Subscripts interpretation

For C nozzles, r_{po} is the ratio between the output and input pressure. For CD nozzles r_{po} is the ratio between throat and input pressure. See group of Formulas 3.5.

Formulas 3.5. Pressure ratios of Saint Venant - Wantzel equation

For CD nozzle: $r_{po} = r_{pt} = \dfrac{p_t}{p_1}$ For C nozzle: $r_{po} = r_{pn} = \dfrac{p_2}{p_1}$

As flow in CD nozzles is choked r_{pt} equals the critical pressure ratio r_{pc}.

A useful mass flow formula emerges from the Saint Venant - Wantzel equation, which will be discussed in this chapter. It must be noted that Formula 3.4 demonstrates that flow velocity depends only on one state property: the input temperature.

However, there is a caveat in the Saint Venant–Wantzel equation application: its mathematical expression assumes that the outgoing jet discharges in a uniform pressure environment, but this is not correct. The pressure seen by the jet is higher than the discharge pressure. This is due to the flow accommodation to its new environment. The consequence, is that the jet velocity is lower than in the ideal case of constant discharge pressure.

3.3. Speed of sound and critical ratios

It has been previously said that the fluid undergoes a strong acceleration in the converging part of a CD nozzle, which can take the flow velocity from a negligible initial velocity ($V_1 \cong 0$) to the speed of sound at the throat. When the flow velocity at the throat equals the speed of sound, the flow is choked and the nozzle is operating under critical conditions. In that case the speed of sound is given by:

Formula 3.6. Speed of sound

$$V_s = \sqrt[2]{g \cdot k \cdot J \cdot R \cdot T_t} \quad \left|\frac{m}{s}\right|$$

Similarly, as in the Saint Venant–Wantzel equation, the Formula 3.6 demonstrates that the speed of sound in a fluid depends only on one state property: the temperature. Also, an alternative form of this formula is to replace the $J \cdot R \cdot T_t$ product by the $p_t \cdot v_t$ product, as defined by the equation of state.

The critical ratios of pressure, temperature and specific volume at the throat, are numbers that allow to calculate the variables of state at the throat as a function of their homologous at the nozzle inlet. They have been defined before and are derived in section 8.4.

The resulting formulas of this procedure have a great importance in the design and evaluation of a nozzle for they allow the critical properties calculation, regardless of the discharge pressure. The critical ratios values, is only dependent on the specific heat ratio k. Table 1.1 shows k values for different fluids used in turbines.

The critical ratios formulas are as follows.

Formula 3.7. Critical pressure ratio

$$r_{pc} = \frac{p_{tc}}{p_1} = \left(\frac{2}{k+1}\right)^{\frac{k}{k-1}}$$

Formula 3.8. Critical temperature ratio

$$r_{tc} = \frac{T_c}{T_1} = \frac{2}{k+1}$$

Formula 3.9. Critical specific volume ratio

$$r_{vc} = \frac{v_c}{v_1} = \left(\frac{k+1}{2}\right)^{\frac{1}{k-1}}$$

The subscript c means critical values at the throat, caused by the flow choking. See Section 1.10. The values yielded by these formulas are in Table 1.1.

Formulas have been developed for ideal gases, but are equally applicable to superheated, saturated and wet steam of high quality. Although the results are not exact, they are considered a good approximation. For example, for superheated steam k = 1.3 is usually adopted. Saturated or wet steam k values must be calculated with the Zeuner Formula 3.10.

Formula 3.10. Specific heat ratio of saturated and wet steam

$$k = 1.035 + 0.1 \cdot x$$

The following example will help to understand the application of the three critical relationships given by Formulas 3.7, 3.8 and 3.9.

3.4. Example of critical properties calculation

Let there be a nozzle having a fluid with inlet pressure of 15 kg/cm² a, at a temperature of 600°K. The working fluid is combustion gas.

Determine the value of pressure, temperature and specific volume at the throat, under choked flow conditions.

○ Critical ratios:

The critical ratios are in Table 1.1, in the combustion gas line.

$r_{pc} = 0.5404$ $r_{tc} = 1.5885$ $r_{vc} = 0.8584$

○ Throat pressure:

$p_t = 0.5404 \times 15 = 8{,}106 \text{ kg} / \text{cm2}$

o   Throat temperature:

$T_t = 0.8584 \times 600 = 514.04 \,°\,K$

o   Input and throat specific volumes calculation:

$$v_1 = \frac{J \cdot R \cdot T_1}{p_1} = \frac{102 \times 0.287 \times 600}{15 \times 10^4} = 0.1171 \left| \frac{m^3}{kg} \right|$$

$$v_t = 0.1171 \times 1.5885 = 0.1860 \quad \left| \frac{m^3}{kg} \right|$$

3.5. Throat and outlet velocities

The Saint Venant–Wantzel equation is used to calculate the throat velocity in CD nozzles or the discharge velocity in C nozzles. Of course, this equation always returns the speed of sound in a CD nozzle because flow is choked at its throat.

By replacing Formula 3.7 in Formula 3.4, velocity of speed is returned, which is the maximal speed at the throat.

Formula 3.11 Formula of throat velocity=speed of sound

$$V_Ð = a \cdot \sqrt[2]{J \cdot R \cdot T} \qquad \left| \frac{m}{s} \right|$$

Formula 3.11 is simplified by using the following easy procedure to calculate the throat velocity. To this purpose Formula 3.11 is expressed as the product of one constant times the square root of the input temperature. See Formula 3.12.

Formula 3.12. General formula of the
speed of sound at the throat

$$V_t = a_V \cdot \sqrt[2]{J \cdot R \cdot T_1} \qquad \left| \frac{m}{s} \right|$$

Where the throat velocity constant a_V is given by:

Formula 3.13. Throat velocity constant in formula 3.12

$$a_V = \sqrt[2]{2 \cdot g \cdot \frac{k}{k+1}} \qquad \left| \frac{\sqrt[2]{m}}{s} \right|$$

The values of throat velocity constant a_V are tabulated in Table 3.2.

3.6. Example of velocity curves calculation

For this section, refer to file 3.2. Figures 3.2 and 3.3. Velocity and mass flow versus pressure ratio.xlsx.

For the input data and physical constants given in Table 3.1, determine:

o   Throat temperature
o   Speed of sound at the throat
o   Plot of velocity versus pressure ratio r_{pt}

Combustion gas from a gas turbine behaves almost as air because it usually has a very high air to fuel ratio. Hence, values of k and R are very similar to air.

Table 3.1. Input data

Input data			
Property	**Designation**	**Fluid: combustion gas**	
Input pressure	p1	15	kg/cm2 a
Input temperature	T1	873	°K
Area at the throat	At	1.00	cm2
Physical constants			
R		0.331	kJ/kg.°K
J		102	kJ/kg.m
Gravity		9.81	m/s2
k		1.330	

○ Step 1. Throat temperature. Use Formula 3.8.

$$T_t = \frac{2}{1.33+1} \times 873 = 749°K$$

○ Step 2. Maximum velocity or speed of sound at the throat. Use Formula 3.6.

$$V_s = \sqrt[2]{1.33 \times 9.81 \times 102 \times 0.331 \times 749} = 575 m/s$$

○ Step 3. Plot of throat velocity versus pressure ratio. Use Saint Venant - Wantzel equation (Formula 3.4)

For nozzle pressure ratios r_{pn} higher than r_{pc} the plot is calculated with Formula 3.11, which in this case is the following expression.

$$V(r_{pt}) = \sqrt[2]{\frac{2 \times 9.81 \times 1.33}{1.33-1} \left[1 - (r_{pt})^{\frac{1.33-1}{1.33}} \right] \times \sqrt[2]{102 \times 0.331 \times 873}} = 1,526 \times \sqrt[2]{1 - r_{pt}^{0.248}}$$

○ From Table 1.1 the critical ratio for combustion gas is r_{pc} = 0.5404. Hence, velocity will be equal to the speed of sound for any value lower than p_{tc} = 0.5404×15 = 8.1 kg/cm².

The resulting velocity curve is in Figure 3.2. It must be noted the curve discontinuity at pressure ratio equal to its critical value.

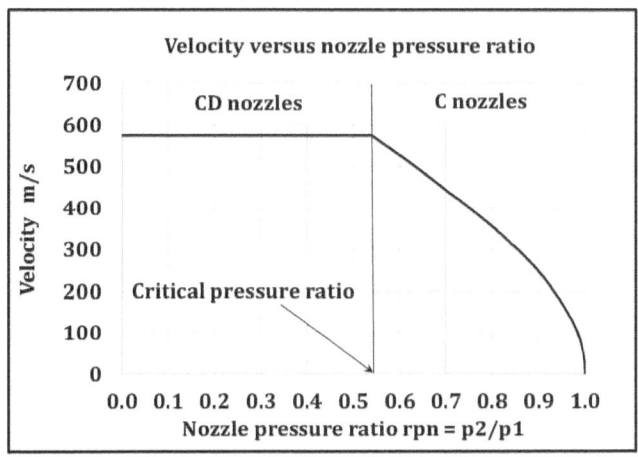

Figure 3.2. Flow velocity curve versus pressure ratio

The results obtained in this example are significant for the assessment of a nozzle performance. It must be noted that the velocity curve was calculated with only two thermodynamic properties: the input pressure and temperature. These properties are easy to obtain at the field, when evaluating nozzle performance. These results could be used for nozzle benchmarking purposes.

3.7. Mass flow in a choked flow

Any fluid mass flow, confined in a pipe or nozzle or any other type of enclosure, is derived from Formula 1.5. Principle of mass continuity The resulting formula is applicable to any section of the nozzle and according to the principle of mass continuity, the result is always the same. Therefore, mass flow at the outlet section is calculated with the following general formula.

Formula 3.14. Mass flow

$$G = \frac{A_x \cdot V_x}{V_x} \qquad \left|\frac{kg}{s}\right|$$

In Section 8.5 the mass flow formula, applicable to gas and steam, is derived. The resulting expression for any section in the nozzle is the following:

Formula 3.15. Mass flow versus nozzle pressure ratio

$$G(r_{po}) = \sqrt[2]{\frac{2 \cdot k \cdot g}{k-1} \cdot \left[(r_{po})^{\frac{2}{k}} - (r_{po})^{\frac{k+1}{k}}\right]} \cdot \sqrt[2]{\frac{p_1}{v_1}} \cdot A_o \qquad \left|\frac{kg}{s}\right|$$

Where A_o is the orifice area.

This formula indicates that the transport capacity depends on the pressure ratio, the square root of p_1/v_1 ratio and the orifice area A_o.

Following with the example of section 3.6 the mass flow is calculated with Formula 3.15. This formula requires to know the input specific volume. This is calculated with the equation of state as follows.

○ Specific volume at the inlet:

$$v_1 = \frac{J.R \cdot T_1}{p_1} = \frac{102 \times 0.331 \times 873}{15 \times 10^4} = 0.196 m^3/kg$$

○ The formula of mass flow curve is:

$$G\left(r_{pn}\right) = \sqrt[2]{\frac{2 \times 1.33 \times 9.81}{1.33 - 1} \cdot \left[\left(r_{pn}\right)^{\frac{2}{1.33}} - \left(r_{pn}\right)^{\frac{1.33+1}{1.33}}\right]}$$

$$\cdot \sqrt[2]{\frac{15 \times 10^4}{0.196}} \cdot 1 \times 10^4$$

$$= 0.777 \times \sqrt[2]{r_{pn}^{1.504} - r_{pn}^{1.752}}$$

The plot for this case is in Figure 3.3. The left branch of $G(r_{pn})$ only represents a mathematical existence, but it's invalid for technical applications because the flow is choked.

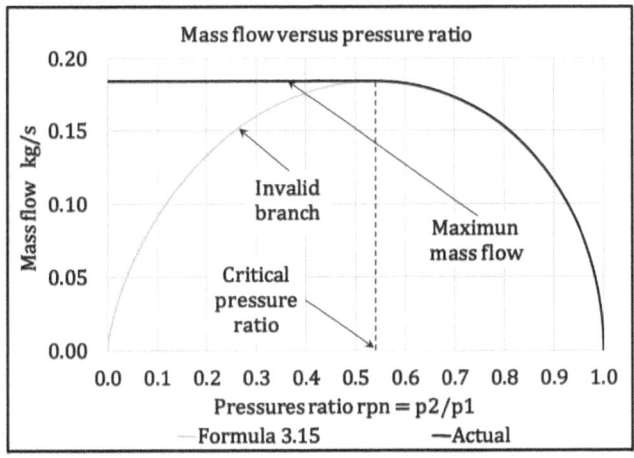

Figure 3.3. Flow mass curve versus pressure ratio

The maximum transport capacity for a choked flow is constant and equal to the maximum mass flow of Formula 3.15. It is calculated with the following formula (See demonstration in section 8.5)

Formula 3.16. Maximum mass flow

$$G_{max} = \left[\sqrt[2]{g \cdot k \cdot \left(\frac{2}{k+1} \right)^{\frac{k+1}{k-1}}} \right] \cdot \sqrt[2]{\frac{p_1}{v_1}} \cdot A_t \qquad \left| \frac{kg}{s} \right|$$

The term between brackets is the mass flow constant a_G, which only depends on k, that is from the fluid specific heats. The resulting G_{max} value is represented by the horizontal portion of the curve named actual mass flow in Figure 3.3.

Formula 3.17. Mass flow constant

$$a_G = \sqrt[2]{g \cdot k \cdot \left(\frac{2}{k+1} \right)^{\frac{k+1}{k-1}}} \qquad \left| \frac{\sqrt[2]{m}}{s} \right|$$

The maximum value of G is calculated with the following formula, derived from Formulas 3.16 and 3.17.

Formula 3.18. Maximum mass flow. Compacted formula

$$G_{max} = a_G \cdot \sqrt[2]{\frac{p_1}{v_1}} \cdot A_t \qquad \left| \frac{kg}{s} \right|$$

a_G values are listed in Table 3-2 for different working substances.

Table 3.2. Velocity and mass flow constants a_v and a_G

Formulas 3.13 and 3.17

Velocity and mass flow constants				
Fluid	Steam title	k	aV |m^0.5/s|	aG |m^0.5/s|
Wet steam	0.800	1.115	3.216	1.978
Wet steam	0.850	1.120	3.220	1.981
Wet steam	0.900	1.125	3.223	1.984
Wet steam	0.950	1.130	3.226	1.988
Saturated steam	1.000	1.135	3.230	1.991
Superheated steam		1.300	3.330	2.090
Combustion gases		1.330	3.347	2.107
Air		1.400	3.383	2.145
Monoatomic gases		1.667	3.502	2.275

For this table, refer to file 3.3. Table 3.2. Velocity and mass flow constants.xlsx.

3.8. Example of a critical velocity and mass flow calculation

A conical nozzle operates with a flue gas at a pressure of 18 kg/cm2 a, and a temperature of 1,000°C. As was said before, the combustion gas has a very high air/fuel ratio, hence, its properties are similar to air. If a CD nozzle is necessary, use a throat diameter of 2 cm and a discharge diameter of 4.5 cm.

Determine:

Critical speed and maximum mass flow for two different discharge pressures:

- Case a) $p_2 = 4$ kg/cm2 a. See subsection 3.8.1.
- Case b) $p_2 = 14$ kg/cm2 a. See subsection 3.8.2.

In each case determine whether the nozzle must be C or CD.

If the nozzle type proves to be CD calculate the following properties.

- o Maximum mass flow
- o Specific volume expansion: $X_v = v_2/v_t$
- o Velocity expansion: $X_v = V_2/V_t$.
- o Geometric expansion: $X_t = A_2/A_t$
- o Compare X_v with X_t.
- o Calculate the Mach number of the discharge velocity.

3.8.1. Case a) p_2 = 4 kg / cm2

- o Nozzle pressure ratio:

$$r_{pn} = \frac{4}{18} - 0.222$$

- o Critical pressure ratio for combustion gases from Table 1.1: r_{pc} = 0.5404
- o As $r_{pn} < r_{pc}$, the flow is choked. Hence, the nozzle must be CD type. Formulas 3.6 and 3.16are the appropriate to calculate V_t and G_{max} respectively.
- o Specific volume calculation at the nozzle inlet: R_{cg} = 0.331 kJ / (kg. ° K)

$$V_1 = \frac{J \cdot R \cdot T_1}{P_1} = \frac{102 \times 0.331 \times 1{,}273}{18 \times 10^4} = 0.239 \text{ m}^3/\text{kg}$$

- o Critical velocity:
  Values of the constants a_v and a_G are read in Table 1.1, in the combustion gases row.

  a_v = 3,347

$$T_t = r_{tc} \cdot T_1 = 0.8584 \times 1{,}273 = 1{,}093°K$$

$$V_t = a_v \cdot \sqrt[2]{J \cdot R \cdot T_1} = 3.347 \times \sqrt[2]{102 \times 0.331 \times 1{,}273} = 693 \text{ m/s}$$

o   Throat area:

$$A_t = \frac{\pi \cdot 2^2}{4} = 2.14 \text{ cm}^2$$

o   Maximum mass flow:

$$a_G = 2.107$$

$$A_t = \frac{\pi \times \left(2 \times 10^{-2}\right)^2}{4} = 3.142 \times 10^{-4} \text{ cm}^2$$

$$G_{max} = 2.107 \times \sqrt[2]{\frac{18 \times 10^4}{0.239}} \times 3.142 \times 10^{-4} = 0.575 \text{ kg/s}$$

o   Specific volume expansion ratio:

From Table 1.1: k = 1.33

$$V_t = \frac{V_1}{r_{cp}^{\left(\frac{1}{k}\right)}} = \frac{0.239}{0.5404^{\frac{1}{1.33}}} = 0.380 \text{m3/kg}$$

$$V_2 = V_1 \cdot \left(\frac{p_1}{p_2}\right)^{\frac{1}{k}} = 0.239 \times \left(\frac{18}{4}\right)^{\frac{1}{1.33}} = 0.740 \text{m3/kg}$$

o   Expansion ratio in the divergent part:

$$X_v = \frac{V_2}{V_1} = \frac{0.740}{0.380} = 1.947$$

Conclusion: the gas is expanded 95% in the divergent part, which strongly contributes to the velocity growth.

○ Velocity expansion ratio:

$$V_2 = 2\sqrt{\frac{2 \times 9.81 \times 1.33}{1.33 - 1}\left[1 - \left(\frac{4}{18}\right)^{\frac{1.33-1}{1.33}}\right]}$$
$$\times \sqrt[2]{102 \times 0.331 \times 1,273} = 1,029 \, m/s$$

$$X_v = \frac{V_2}{V_c} = \frac{1,029}{646} = 1.593$$

Conclusion: In the divergent part the velocity is increased by 59%.

○ Geometric expansion:

$$X_g = \frac{A_2}{A_t} = \frac{D_2^2}{D_t^2} = \left(\frac{4.5}{2.0}\right)^2 = 5.06$$

○ Specific volume expansion over velocity expansion:

$$r_x = \frac{X_v}{X_V} = \frac{1.947}{1.593} = 1.22$$

Therefore, the specific volume expanded by 22% over the velocity growth.

The divergent part has a significant geometric expansion (5.06 times), which would usually produce, under other circumstances, a velocity drop, just like a diffuser. However, as the flow is supersonic and choked, this drop doesn't happen. Furthermore, the velocity, in this case, increases by 59.3%.

o   Mach number of the discharge flow

Discharge speed of sound:

$$V_{2s} = \sqrt[2]{1.33 \times 9.81 \times 4 \times 10^4 \times 0.740} = 621 m/s$$

Note: speed of sound is different than velocity at the throat because of the thermodynamic state differences between the throat and the discharge.

Discharge Mach number:

$$M_2 = \frac{V_2}{V_{2s}} = \frac{1,029}{621} = 1.657$$

The discharge velocity is 66% higher than the speed of sound.

3.8.2. Case b) p_2 = 14 kg/cm2 a

o   Nozzle pressure ratio:

$$r_p = \frac{14}{18} = 0.778$$

As $r_{pn} > r_{pc}$ the flow is not choked. The nozzle should be type C.

o   Discharge flow velocity. Formula 3.1.

$$V_2 = \sqrt[2]{\frac{2\times9.8\times1.33}{1.33-1}\left[1-\left(0.778\right)^{\frac{1.33-1}{1.33}}\right]} \times \sqrt[2]{102\times0.331\times1,273} = 486m/s$$

- Mass flow

$$G = \sqrt[2]{\frac{2\times1.33\times9.81}{1.33-1}\cdot\left[\left(0.778\right)^{\frac{2}{1.33}}-\left(0.778\right)^{\frac{1.33+1}{1.33}}\right]} \times 3.142\times10^{-4}\times\sqrt[2]{\frac{18\times10^4}{0.239}} = 0.494\ kg/s$$

The above results suggest the following conclusions:

- The discharge velocity ratio of the CD nozzle over the C nozzle is equal to: $\frac{1,029}{486} = 2.18$. The square of this number is the kinetic energy ratio between the CD nozzle and the C nozzle, which is: $2.27^2 = 4.48$. It means that, the outlet kinetic energy of the CD nozzle is 4.48 times higher than that of the C nozzle.
- This lower discharge kinetic energy of the C nozzle respect to the CD nozzle, doesn't mean that C nozzles have a poorer performance, because the required outflow velocity depends on the turbine construction and a high kinetic energy is not always desirable for the turbine performance. Furthermore, in many turbines the kinetic energy of the flow is fractioned between more than one blades wheel, because otherwise the wheel could reach unacceptable velocities.
- As for the mass flow, the ratio of the CD nozzle flow to the C nozzle flow equals:

$$\frac{G_c}{G_{CD}} = \frac{0.575}{0.494} = 1.16$$

This number demonstrates that the CD nozzle has 16% higher transport capacity than the C nozzle. However, this is not a universal statement. It's only valid for these two nozzles, which have equal throat diameters and the same inlet thermodynamic conditions.

○ Mach number of the discharge flow:
○ Discharge specific volume:

$$V_2 = V_1 \cdot \left(\frac{p_1}{p_2}\right)^{\frac{1}{k}} = 0.239 \cdot \left(\frac{18}{14}\right)^{\frac{1}{1.33}} = 0.289 \, m3 \, / \, kg$$

○ Discharge temperature:

$$T_2 = \frac{p_2 \cdot V_2}{J \cdot R} = \frac{14 \times 10^4 \times 0.250}{102 \times 0.331} = 1,037°K$$

○ Speed of sound:

$$V_s = \sqrt[2]{1.33 \times 9.81 \times 102 \times 0.331 \times 1,037} = 676 \, m \, / \, s$$

○ Mach number

$$M_2 = \frac{V_2}{V_s} = \frac{486}{676} = 0.719$$

This value equals 43.3% of the CD nozzle Mach number. Converging nozzles operate with Mach numbers smaller than 1 and close to unity.

Chapter 4

Flow perturbations: Friction, Turbulence and Shock Waves

4.1. Friction in nozzles

The Darcy – Weisbach formula returns the kinetic energy loss in a pipe caused by friction. If this energy loss is multiplied by the specific gravity, the result is the pipe pressure loss produced by the same cause. This last modality of the Darcy – Weisbach formula application, is specially used for long pipeline design, where pressure loss is an important impact that must be minimized. In this section, the application of the Darcy – Weisbach formula to nozzles purports to calculate the heat q_f released by friction between the fluid and the nozzle's inner walls.

The Darcy – Weisbach formula considers that the kinetic energy loss of a flow in a pipe, depends on:

○ The pipe length over the diameter ratio. As this ratio increases, the total heat released by friction also escalates.
○ The pipe roughness and the flow's Reynolds number determine the energy loss by friction and their consolidated action is reduced to one coefficient, smaller than 1, known as friction factor.

There does not exist a simple derivation of formulas to calculate the friction factor. Empirical expressions have been derived, which successfully provide the results according to the experience. All the empirical formulas arrive at the same conclusion: friction factor is higher as nozzle roughness increases and lower as flow Reynolds number also increases.

The kinetic energy loss of a nozzle's flow, is calculated with the Darcy – Weisbach formula.

Formula 4.1. Energy loss by friction
with Darcy - Weisbach formula

$$q_f = f\left(\varepsilon_r, Re\right) \cdot \frac{L_t}{D_{avg}} \cdot E_{k2\,ideal} \qquad \left|\frac{kJ}{kg}\right|$$

Where:

q_f is the kinetic energy loss or heat released by friction, in kJ/kg.

L_t is the nozzle total length in m or cm.

D_{avg} is the nozzle average diameter, expressed in the same unit used for L_t.

$E_{k2\,ideal}$ is the ideal kinetic energy of the flow at the nozzle outlet, in kJ/kg.

$f(\varepsilon_r, R_e)$ is the friction factor

ε_r is the relative roughness of the inner walls = ε / D_{avg}

ε is the absolute roughness, expressed in the same units than D_{avg}

R_e is the Reynolds number as defined in Formula 4.2.

The Reynolds number indicates whether a flow is turbulent. Mathematically it is equal to inertial forces over viscous forces ratio. As Reynolds number is higher, turbulence is also higher.

Reynolds number of gas is impacted by temperature because gas viscosity grows with temperature.

The Reynolds number formula flow for pipes is as follows.

Formula 4.2. Reynolds number

$$R_e = \frac{\rho \cdot V \cdot D}{\mu}$$

Where:

ρ: fluid density. Mass kg/m³
V: flow velocity. m/s
D: nozzle diameter at the outlet. m
μ: dynamic viscosity of the fluid. Mass kg/[m·s]

Formula of the dynamic viscosity is:

Formula 4.3. Dynamic viscosity

$$\mu = \mu_0 \cdot \left(\frac{T_x}{T_{ref}}\right)^{1.5} \cdot \left(\frac{T_{ref} + S_u}{T_x + S_u}\right) \qquad \left|\frac{\text{mass kg}}{\text{m} \cdot \text{s}}\right|$$

Where:

μ_0: Air viscosity at reference temperature = 1.716×10^{-5} mass kg/|m.s|
T_{ref}: Reference temperature = 323°K
S_u: Sutherland constant = 110°K
T_x: Air temperature in the nozzle

4.1.1. The Moody diagram

Friction factor is calculated with the Moody diagram, which is reproduced in Figure 4.1, or with the following empirical formula, also derived by Moody.

Formula 4.4. Moody formula for friction factor

$$f = 0.0055 \times \left[1 + \left(2 \times 10^4 \cdot \varepsilon_r + \frac{10^6}{R_e} \right)^{\frac{1}{3}} \right]$$

This formula returns acceptable values of friction factor in the typical nozzle's roughness range, which is from 0.001 to 0.006. There are other empirical expressions of function $f(\varepsilon_r, Re)$, like the Colebrook formula, which may be consulted in the excellent available bibliography about Fluid Mechanics and also in Wikipedia, page: Darcy Friction Factor. Under the Bibliography title a list of recommended books is produced. However, there exist lots of scientific and technical books that can be used to learn the theoretical basis of all formulas used in this chapter.

The Moody diagram presents four regions (See Figure 4.1).

o Laminar flow. This region corresponds to very low Reynolds number (<2,500) and high friction factors. The friction factor is inversely proportional to the Reynolds number. The usual formula for this region if $f = 64/Re$. High viscosity fluids are usually in this region; however, nozzles do not operate in this region.
o Transition region. This is between $Re = 2,000$ and $Re = 4,000$. This is a mix of laminar and turbulent regime, where friction factor depends on Reynolds number and the inner wall roughness. Nonetheless, no reliable friction factor formula

exists in this region. Nozzles do not usually operate in this region.

○ Turbulent region. This is located between the smooth pipe and the complete turbulence curves. Friction factor of this region depends on both, the Reynolds number and the relative roughness. Formula 4.4 is applicable.

○ Complete turbulent region. This region is beyond the complete turbulent curve (dotted line). In this region friction factor only depends on the relative roughness. Horizontal curves indicate thatin this region Reynolds number does not affect friction factor. Formula 4.4 is applicable.

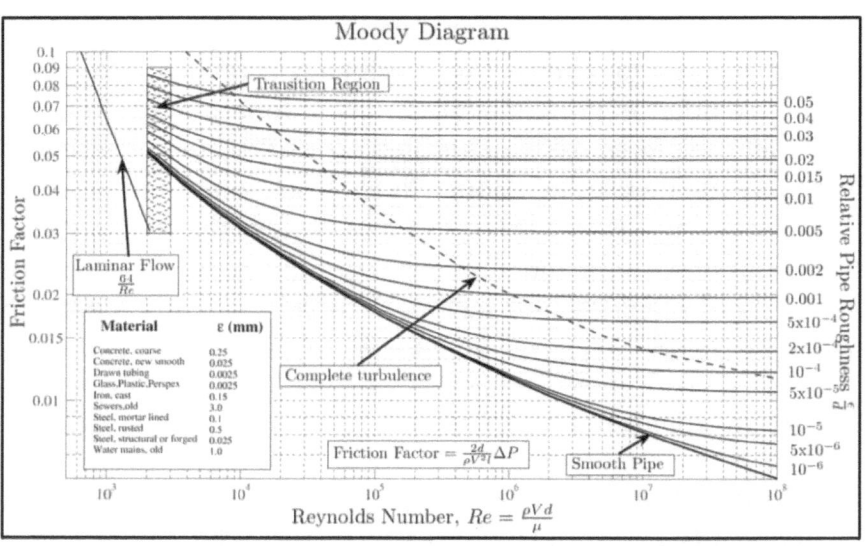

Figure 4.1. Moody diagram

As nozzles usually operate with high Reynolds number (typically higher than 60,000) and have low relative roughness (typically between 0.005 and 0.050), friction factor calculation comes from both the turbulent and the complete turbulent regions. It's important to remark that as Reynolds number increases in the first three regions (laminar, transition and turbulent), friction factor decreases.

4.1.2. Velocity coefficient formula

For this section refer to file 4.4. Figure 4.2. Velocity coefficient and Darcy formula.xlsx.

The velocity coefficient is calculated on the basis of friction factor. As Formula 2.10 and Formula 4.1 represent both the same q_f value, they are equated to derive the velocity coefficient formula versus friction factor and L_t/D_{avg} ratio. The result is the following expression.

Formula 4.5. Friction heat q_f (reheat)

$$q_f = E_{k2ideal} - E_{k2} = f\left(\varepsilon_r, Re\right) \cdot \frac{L_t}{D_{avg}} \cdot E_{k2ideal} \qquad \left|\frac{kJ}{kg}\right|$$

Where:

Formula 4.6. Actual kinetic energy at the outlet

$$E_{k2} = \varphi^2 \cdot E_{k2ideal} \qquad \left|\frac{kJ}{kg}\right|$$

From Formula 4.5 and 4-6 arises the velocity coefficient versus friction factorand the nozzle L_t/D_{avg} ratio.

Formula 4.7. Velocity coefficient versus friction factor

$$\varphi = \sqrt[2]{1 - f\left(\varepsilon_r, Re\right) \cdot \frac{L_t}{D_{avg}}}$$

Formula 4.7 is visualized in Figure 4.2, for six different values of friction factor f.

The plotted curves are in the 0.03 to 0.08 friction factor range, which exceeds a typical range of gas and steam nozzles.

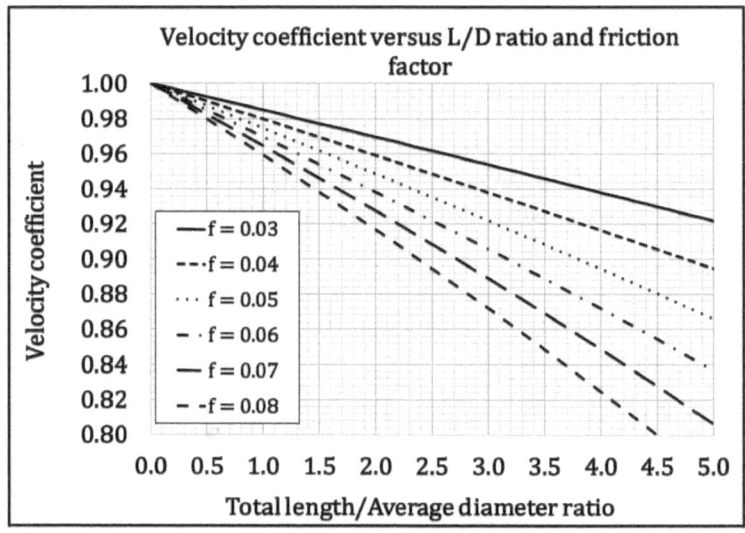

Figure 4.2. Velocity coefficient versus L/D ratio and friction factor. Darcy – Weisbach formula

Sometimes, the following formula is used:

Formula 4.8. Velocity coefficient. Approximate formula

$$\varphi = \sqrt[2]{1 - 0.03 \times \frac{L_t}{D_{avg}}}$$

This expression assumes that friction factor is 0.03. This value corresponds to roughness coefficients between 0.005 and 0.007 and Reynolds numbers higher than 50,000. However, it is not recommended to use this formula if a roughness reliable value is available.

The conclusion of Formula 4.7 is that long nozzles of small diameter have lower velocity coefficients than short nozzles of large diameter and that the nozzle roughness has a higher impact on friction factor than Reynolds number in the turbulent and complete turbulent regions. This will be demonstrated with an example in section 4.2.

Formula 4.7 reunites the thermodynamic considerations about velocity and efficiency (φ and φ^2 respectively) with two very important nozzle constructive aspects, which are roughness and L/D ratio.

4.2. Example of nozzle efficiency sensitivity versus relative roughness

See file 4.2. Table 4.1. Efficiency sensitivity to roughness.xlsx.

Let it be two identical CD nozzles for superheated steam, but with different relative roughness. Determine each nozzle efficiency assuming that the convergent part does not participate in the nozzle loss. Therefore, the input data for this calculation are:

○ The nozzle diameter is the average diameter of the divergent part.
○ The nozzle L/D ratio is calculated with the length and average diameter of the divergent part.
○ The relative roughness of nozzle 1 is four times lower than nozzle 2 (0.002 and 0.008 respectively).
○ Both nozzles have the same Reynolds number.

Table 4.1. Example of efficiency sensitivity to roughness

Input data		
Property	**Nozzle 1**	**Nozzle 2**
Relative roughness ε_r	0.002	0.008
Length/Average diameter	2.675	2.675
Reynolds number R_e	130,326	130,326
Formulas and results		
Property	**Formula**	**Nozzle 1**	**Nozzle 2**
Friction factor f	See Formula 4.4	0.0254	0.0358
Velocity coefficient φ Formula 4.8	$\varphi = \sqrt[2]{1 - f\left(\varepsilon_r, \mathrm{Re}\right) \cdot \dfrac{L_t}{D_{avg}}}$	0.965	0.951
Nozzle efficiency % Formula 2.9	$e_n = \varphi^2 \cdot 100$	93.2%	90.4%

The aftermath of these calculations is that a 300% increase of the relative roughness produces a 41% increase of friction factor and consequently, 2.8% reduction of the nozzle efficiency. Furthermore, friction factor % increase is exactly the same % as the friction heat released (q_f) by the nozzle and incorporated to the discharge flow.

What happens if the Reynolds number grows to double the initial value? In this case the nozzle is operating on a very flat region of the Moody diagram. That means that a strong Reynolds number increase, does not produce any significant change of friction factor, therefore, the efficiency is not affected.

4.2. Example of nozzle efficiency calculation

For this section refer to file 4.3. Tables 4.2 and 4.3. Nozzle efficiency calculation.xlsx.

The fundament of this example is that nozzle geometry and its thermodynamic state variables are known. SeeTable 4.1.

Table 4.2. Input data: nozzle geometry and thermodynamic state

Fluid	Air			
Nozzle type	CD			
Relative roughness	0.009			
Acceleration of gravity	9.81	m/s2		
Property	**Inlet**	**Throat**	**Outlet**	**Units**
Diameter	5.10	2.58	2.88	cm
Length	1.22	0.00	2.88	cm
Temperature	1,300	1,083	875	°K
Specific gravity	3.96	2.51	1.47	kg/m3
Velocity	0	760	1,044	m/s
Dynamic viscosity data				
At reference temperature	1.716E-05	mass kg/(m.s)		
Reference temperature	323	°		
Sutherland constant	110	°		

Determine the nozzle efficiency with data of Table 4.2, on the following assumptions:

o   Friction loss in the convergent part is not negligible

o Nozzle equivalent length equals total length that is convergent plus divergent part lengths.
o Average diameter is weighted by each part length, with the following formula:

Formula 4.9. Nozzle equivalent diameter for
Reynolds and Darcy – Wantzel formulas

$$D_{avg} = \frac{L_1}{L_t} \cdot \frac{D_1 + D_t}{2} + \frac{L_2}{L_t} \cdot \frac{D_t + D_2}{2} \qquad |cm|$$

o Reynolds number is the average of the Reynolds number at the throat and the outlet.

Formulas, calculation sequence and results are in Table 4.3.

It's important to note that this procedure is an approximation because equivalent length, diameter, viscosity and Reynolds are average values. A more precise calculation is obtained with a dedicated CFD software. However, for practical purposes, results of Table 4.2 and 4-3 are acceptable.

Table 4.3. Nozzle efficiency calculation		
Equivalent length. cm	$L_{eq} = L_1 + L_2$	4.10
Average diameter. cm	$D_{avg} = \frac{L_1}{L_t} \cdot \frac{D_1 + D_t}{2} + \frac{L_2}{L_t} \cdot \frac{D_t + 2}{2}$	3.06
L/D ratio	$\dfrac{L_{eq}}{D_{avg}}$	1.339

Property	Formula	Inlet	Throat	Outlet
Dynamic viscosity |mass kg/ (m·s)|	$\mu = \mu_0 \cdot \left(\dfrac{T_x}{T_{ref}}\right)^{1.5}$ $\cdot \left(\dfrac{T_{ref}+S_u}{T_x+S_u}\right)$	$4.26 \cdot 10^{-5}$	$3.83 \cdot 10^{-5}$	$3.36 \cdot 10^{-5}$
Reynolds number	$R_e = \dfrac{\rho \cdot V \cdot D}{i}$	0	131,362	133,994
Equivalent Reynolds	$R_{eq} = \dfrac{\left(R_{et}+R_{e2}\right)}{2}$			132,678
Friction factor	$f = 0.0055 \times \left[1+\left(2\times 10^4 \cdot \varepsilon_r + \dfrac{10^6}{R_{eq}}\right)^{\frac{1}{3}}\right]$			0.0370
Velocity coefficient	$\varphi = \sqrt[2]{1 - f \cdot \dfrac{L_t}{D_{avg}}}$			0.975
Nozzle efficiency	$e_n = \varphi^2 \cdot 100$			95.06%

4.4. Example of turbulence assessment in conical nozzle

For this section refer to file 4.5. Figures 4.3 and 4.4. Reynolds assessment. xlsx.

Nozzle turbulence is not easy to calculate exactly, nonetheless, some assumptions can be done to simplify the turbulence assessment, at least for comparison purposes with other similar nozzles. This assessment is done by assuming that it is valid to calculate the Reynolds number with the same formula used for pipes.

Determine the distribution plot of the Reynolds number in an air CD nozzle, for different discharge pressures. This example is in file 4.4. Figures 4.3 and 4.4. Reynolds assessment.xlsx.

The calculation procedure is detailed below for one section: the outlet. All other sections require the same input data.

o Input data:
o Specific gravity: δ_2 = 2.281 kg / m3
o Velocity: V_2 = 833 m / s
o Outlet diameter: D_2 = 3.70 cm
o Dynamic viscosity at T_{ref}: μ_2 = 1.716 x 10^{-5} kg/|m· s|
o T_2 = 1,035 °K
o Roughness = 0
o First step. Refer the dynamic viscosity to the actual temperature. This is done with the Sutherland Formula 4.3.

μ = 1.716 x 10^{-5} × $(1,035/323)^{1.5}$ × (323 + 110) / (1,035 + 110) = 3.755×10^{-5} kg / |m.s|

o Second step. Calculate the Reynolds number with Formula 4.2 at the outlet.

Re = [(2.281/9.81) × 833 × 0.037)] / 4.255×10^{-5} = 192,272

As roughness is zero, friction factor is produced only by turbulence. Its value is 0.016. See Moody diagram in Figure 4.1, the smooth pipe curve. If roughness is different to zero friction factor, produced by turbulence plus roughness, will be higher than 0.016.

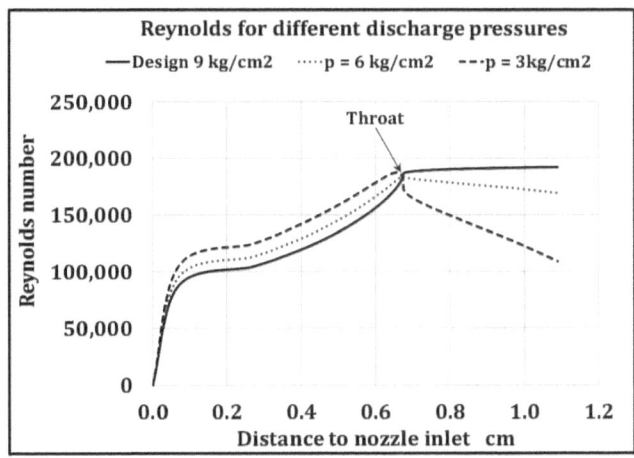

Figure 4.3. Reynolds number distribution
for different discharge pressures

The application of this procedure to other nozzle sections, allows to use a spreadsheet to plot the Reynolds numbers inside the nozzle, versus the abscissa x, as shown in

Figure 4.3. This calculation procedure has to be construed as having a nozzle formed by many straight pipes, of very short length each. In a spreadsheet calculation it's recommended to use at least 50 or more of these elemental pipes.

Figure 4.3 shows the Reynolds numbers curve for the design outlet pressure ($p_2 = 9$ kg/cm2 a) and also the curves for discharge pressures lower than the design pressure. The key properties that drive Reynolds curve behavior are the temperature and the pressure, because the three variables (density, velocity and viscosity) that form the Reynolds formula, are temperature dependent.

Figure 4.3 reveals that in the divergent part of the nozzle, Reynolds number of the design case increases its turbulence to higher values than the other two cases. This is opposite to what happens in the convergent part. As Reynolds number depends directly on the pressure and inversely on to the temperature and there happens a sudden pressure drop at the throat, Reynolds number in the divergent part comes down for pressures below 9 kg/cm2 a.

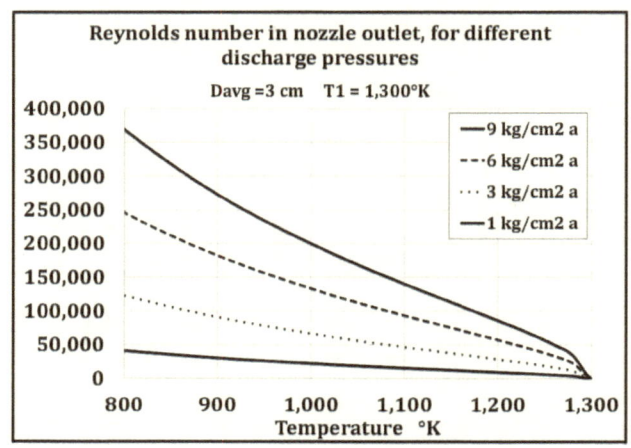

Figure 4.4. Reynolds number versus pressure and temperature

Reynolds number variations inside a nozzle, are mainly due to temperature and pressure changes, because Reynolds formula components (ρ, V, D and μ) are strongly temperature and pressure dependent. The diameter is the exception, although this geometric dimension was determined during the design stage by using properties that are also temperature and pressure dependent.

Note: the plot of Figure 4.4 does not have universal validity. It's only valid for the example of section 4.4.

Each curve corresponds to a discharge pressure value. For T_x = 1,300°K, the velocity is zero because there is not enthalpy difference between the inlet and the outlet. Under this condition no flow can be established in the nozzle.

Curves of Figure 4.4 were obtained by replacing formulas of $\rho(T_x, p_x)$, $V(T_x)$ and $\mu(T_x)$ in Formula 4.2 The result is a non-linear expression of Re versus temperature T_x, but a direct linear relation with the pressure p_x.

Formula 4.10. Detailed form of Reynolds number

$$\mathrm{Re}\left(T_x, p_x\right) = \frac{\left[p_0 \cdot \left(\dfrac{T_0}{T_x}\right) \cdot \left(\dfrac{p_x}{p_0}\right)\right] \times \left[\sqrt[2]{2 \cdot g \cdot J \cdot c_p \cdot \left(T_1 - T_x\right)}\right]}{\left[\mu_{aire} \cdot \left(\dfrac{T_x}{T_{ref}}\right)^{1.5} \cdot \dfrac{T_{ref} + S_u}{T_x + S_u}\right]} \times \left[D\right]$$

Between brackets are the four component formulas of Reynolds number formula. First bracket is the density. This formula follows the ideal gas laws. Second bracket is the velocity and the bracket in the denominator is viscosity. Formula 4.10, although not used in practical applications, explains the importance of temperature T_x and p_x in Reynolds number. With increasing temperature T_x the density and velocity are reduced and the viscosity is increased, which finally produce a Reynolds number reduction.

In summary:

o   In the laminar, transition and turbulent regions of the Moody diagram, friction factor decreases as the Reynolds number increases. Conclusion: higher turbulence creates less friction.
o   Friction factor is practically independent of the Reynolds number in the complete turbulent region (approximately Re > 100,000). Conclusion: higher turbulence does not affect the friction.
o   The nozzle flow turbulence decreases as temperature increases and grows as pressure grows up. Conclusion: pressure and temperature changes affect the turbulence.

4.5. Normal shock waves in CD nozzles

For this section refer to file 4.6. Figures 4.5 and 4.6. Shock wave coefficients.xlsx.

The conceptual definition of a shock wave is in section 1.9. It's recommended to consult it before reading this section.

Subscript y is used for input and z for output variables.

By the application of Physics laws, it is possible to derive formulas to calculate the state variables, velocity and kinetic energy values at the shock wave outlet, versus the values of those same properties at the shock wave inlet. Calculation formulas and the physical principles on which they are based, are given below. The subscript y refers to the shock wave entrance and the subscript z refers to the shock wave output.

The exit Mach number formula is derived from the principle of mass continuity, and the result is as follows.

Formula 4.11. Exit Mach number of a shock wave

$$M_z = \sqrt[2]{\left[\frac{2+(k-1)\cdot M_y^2}{1-k+2\cdot k\cdot M_y^2}\right]}$$

The M_z value given by this formula is always smaller than 1. This indicates that after the shock wave the flow regime becomes subsonic. Therefore, the divergent part between the shock wave and the nozzle exit operates as a diffuser, where pressure increases and velocity goes down. See Figure 4.5.

The exit pressure formula is derived from the principle of conservation of momentum. The result is shown below.

Formula 4.12. Exit pressure of a shock wave

$$p_z = p_y \cdot \left[\frac{1 + k \cdot M_y^2}{1 + k \cdot M_z^2} \right] \qquad \left| \frac{kg}{cm^2} \right|$$

Where p_z is the shock wave output pressure, which is higher than the input p_y pressure. For instance, for $M_y = 3$, the output pressure p_z is ten times the input pressure p_y. See graph of Figure 4.5.

The exit temperature formula is inferred from the principle of conservation of energy, which returns the following expression.

Formula 4.13. Exit temperature of a shock wave

$$T_z = T_y \cdot \left[\frac{2 + (k-1) \cdot M_y^2}{2 + (k-1) \cdot M_z^2} \right] \qquad \left| °K \right|$$

This formula always gives a temperature rise. See graph of Figure 4.5. Consequently, the output enthalpy is increased, and this increase brings adverse consequences for the nozzle efficiency, as was seen before. For instance, for $M_y = 2.6$ the outlet temperature is two times the inlet temperature. See Figure 4.5.

The exit specific volume is derived from the relation between the equations of state at both sides of the shock wave. From this relation the v_z term is cleared and the resulting expression is Formula 4.14.

Formula 4.14. Exit specific volume of a shock wave

$$v_z = v_y \cdot \left[\frac{p_y \cdot T_z}{p_z \cdot T_y} \right] \qquad \left| \frac{m^3}{kg} \right|$$

The exit specific volume v_z is calculated by replacing the results of Formula 4.12 and Formula 4.13 in Formula 4.14. The result is always a lower specific volume than in the shock wave inlet, as shown in Figure 4.5. This means a significant rise of the fluid specific weight between the inlet and the outlet.

The flow velocity, before and after the shock wave, undergoes a significant reduction, as shown in Figure 4.6. The outgoing velocity formula is derived from the speed of sound formula and the Mach number definition. The final expression is Formula 4.15.

Formula 4.15. Exit velocity of a shock wave

$$V_z = V_y \cdot \left[\frac{M_z}{M_y} \cdot \sqrt[2]{\frac{T_z}{T_y}} \right] \qquad \left| \frac{m}{s} \right|$$

The result returned by this formula is the mathematical representation of the velocity loss after the shock wave. The physical aftermath is that the flow is turned from a supersonic into a subsonic regime. Of course, the V_z/V_y ratio is always lower than 1.

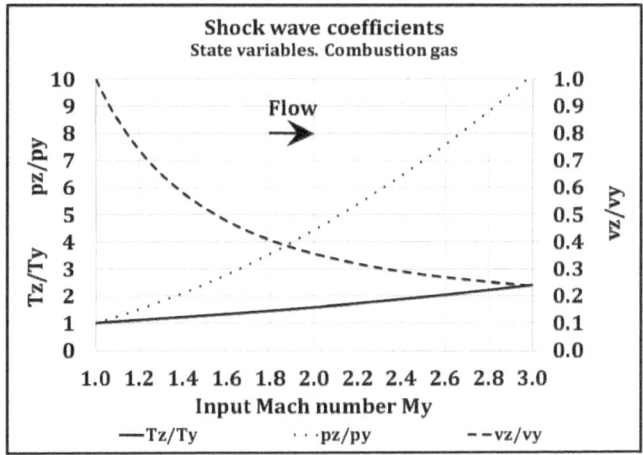

Figure 4.5. Shock wave coefficients. State variables

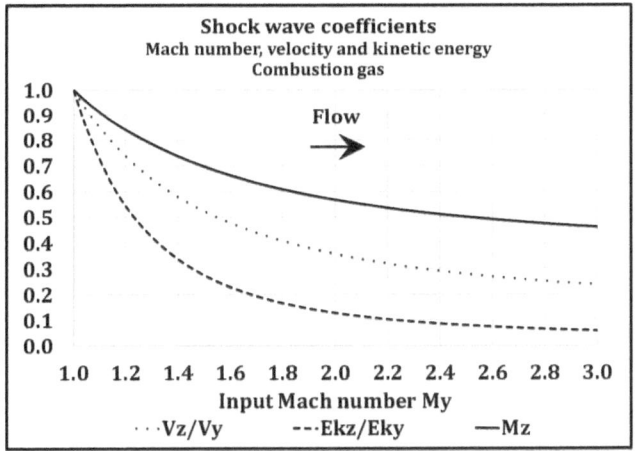

Figure 4.6. Shock wave coefficients. Flow properties

The kinetic energy of the flow is even more affected than the velocity, due to the quadratic law that links the kinetic energy with the velocity.

Formula 4.16. Kinetic energy of the
outgoing flow of a shock wave

$$E_{kz} = E_{ky} \cdot \left[\left(\frac{M_z}{M_y} \right)^2 \cdot \frac{T_z}{T_y} \right] \qquad \left| \frac{kJ}{kg} \right|$$

For example, if the Mach number of the inlet flow is 2, the outgoing Mach number is reduced to 0.567 and the velocity and kinetic energy % losses are equal to 64.4% and 87.3% respectively.

The results of Formula 4.11 to 4.16 are referred in this book as shock wave coefficients. Given the input Mach number My, the outlet values of Mach number, pressure, temperature, specific volume, velocity and kinetic energy, are easily calculated by multiplying the shock wave coefficient by the corresponding input property.

Formula 4.11 to 4-16are only valid to calculate normal shock waves within the nozzle. This validity ends at the discharge section, because waves outside the nozzle are no longer normal but oblique shock waves, which are explained with other formulas.

• In summary:

Formula 4.11 to 4-16 demonstrate that the shock wave increases the pressure and temperature, while the Mach number, specific volume, velocity and kinetic energy are decreased. This behavior indicates that the shock wave inverts the trend that the above flow properties had between the nozzle inlet and the shock wave inlet. As was discussed before, normal shock waves occur because the receiving environment (turbine blades wheel) pressure is higher than the outlet design pressure p_D. This conflict is not resolved by a smooth transition zone, but with an abrupt shock wave. Then this shock wave is the equalizing mechanism of the pressures in conflict.

4.6. Shock wave identification

For this section refer to file 4.7. Figures 4.7 to 4.10. Shock wave plots.xlsx.

Figure 4.7shows different pressure curves, which must be clearly identified to understand the shock waves phenomenon. The extreme points of these curves have been marked with A (Actual discharge pressure), B (Shock wave pressure at the nozzle outlet), D (Design) and T (Throat). The most important characteristics of the pressure behavior are summarized in the following points.

o   Curve T-A of Figure 4.7 shows a critical condition where the flow velocity equals the speed of sound (M = 1) throughout the nozzle. Above this curve the flow regime is subsonic and below it, the flow regime is supersonic. For any discharge pressure equal to, or higher than pressure p_A, no shock waves are produced and the complete divergent part operates as a diffuser, which always works with a subsonic regime.

o   In Figure 4.7, curve T-D shows the design pressure distribution inside the nozzle. This curve is only valid for a discharge pressure equal to the design pressure p_D.

o   Curve T-B of Figure 4.7is the geometric locus of the shock wave pressure p_y established inside the nozzle. It represents the pressure that any internal shock wave reaches. This curve is calculated with Formula 4.11 and 4-12 and in this book is referred to as shock wave pressure curve.

o   The pressure at point B is calculated with Formula 4.17, which for this special point is:

Formula 4.17. Pressure at point B

$$p_B = p_D \cdot \left(\frac{1 + k \cdot M_{yD}^2}{1 + k \cdot M_{zB}^2} \right) \qquad \left| \frac{kg}{cm^2} \right|$$

Where:

p_B: pressure at point B

p_D: discharge pressure of design

M_{yD}: Mach number at the design outlet pressure p_D

M_{zB}: Mach number at the shock wave outlet. This is a shock wave exactly located at the outlet. See Figure 4.10.

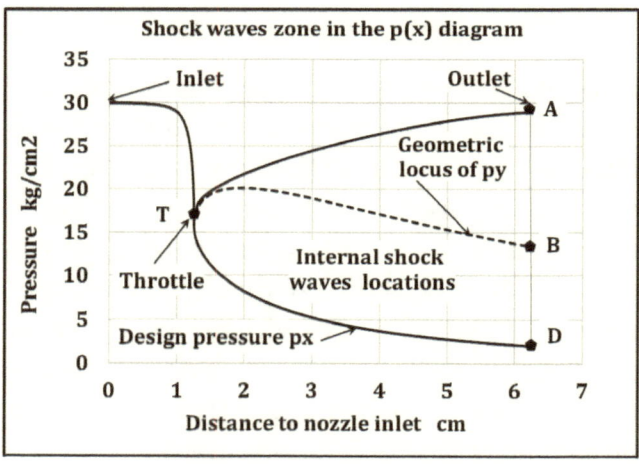

Figure 4.7. Shock wave regions

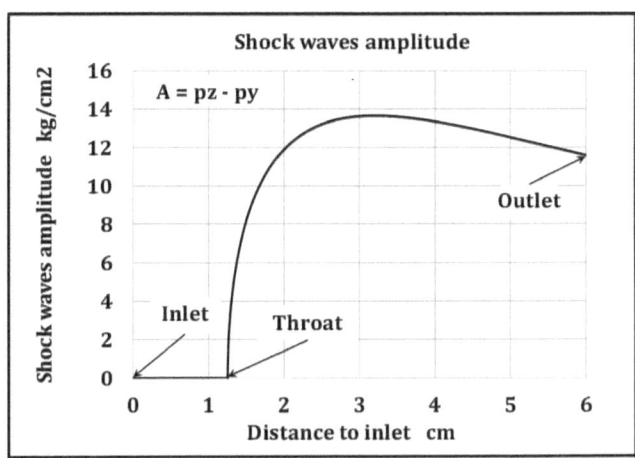

Figure 4.8. Shock wave amplitude

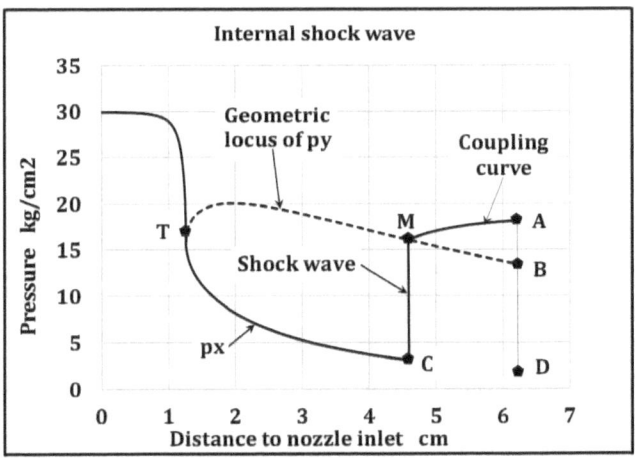

Figure 4.9. Internal shock wave

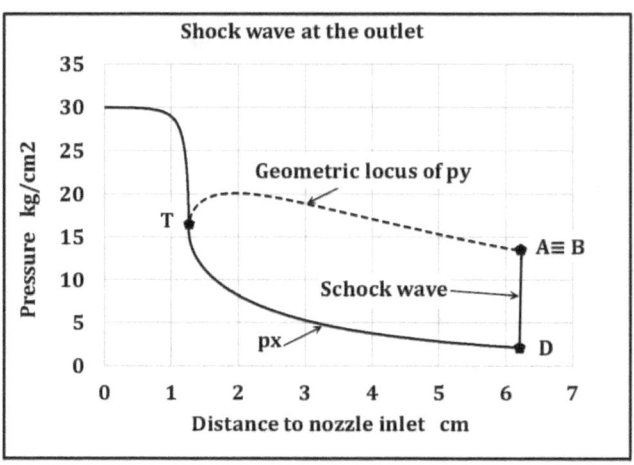

Figure 4.10. Normal shock wave at the outlet

o Figure 4.8 shows the shock wave amplitude curve. The amplitude of a shock wave is the pressure added to the design pressure T-D by a shock wave. Its value is the difference between T-B curve and T-D curve of Figure 4.7. The shock wave amplitude is zero at the throat, and then grows to a maximum, by the middle of the nozzle. After this maximal the amplitude decays to critical pressure p_B. Point B indicates a critical pressure value, below which the shock wave is located beyond the nozzle outlet. The exact location of the shock wave depends on the receiving environment pressure, which might not be exactly equal to the nozzle discharge pressure. However, in these considerations it is supposed that the receiving environmental pressure and the design pressure are the same.

o The C-M-A curve of Figure 4.9 shows the shock wave produced by a discharge pressure at point A higher than the critical pressure of point B. The vertical line C-M represents the shock wave and the curve M-A shows the diffuser behavior between the shock wave and the nozzle outlet. This curve, which is called the coupling curve, smoothly couples the shock wave pressure p_M with the discharge pressure p_A. This shock wave configuration is valid for pressures above the pressure at point B. The exact location of this shock wave is determined by the intersection of the coupling curve M-A with the shock wave pressure curve T-B.

○ Figure 4.9 and 4.10 suggest that the shock wave travels to the nozzle outlet as the receiving environment pressure decreases. This external pressure acts as if it were a plug that pushes against the shock wave. Hence, the higher the external pressure, the closer is the shock wave to the throat. A limit case is the pressure p_t, which locates the shock wave at the throat and makes the entire divergent part of the nozzle to operate as a diffuser. However, it's not possible to state that in this case a shock wave exists, because its input pressure p_y equals its output pressure p_z. Then, both p_y and p_z are equal to the throat pressure p_t. Therefore, no shock wave is established at the throat.

○ Figure 4.10 shows a shock wave exactly located at the nozzle outlet. This happens when the receiving environment pressure equals the point Bpressure, which is located on the shock wave pressure curve (T-B).

○ In summary: normal shock waves inside the nozzle are formed whenever the discharge pressure is between pressures p_A and p_B.

○ For discharge pressures between point B and the design point D, no internal shock waves are formed, but the shock waves formation happens outside the nozzle, provided that $p_2 > p_D$. These external shock waves are oblique to the flow. This operating condition is named overexpanded.

○ For receiving environment pressures lower than that of the design point D, all shock waves are external. These are formed by a series of undulating and oblique shock waves, which equalize the receiving environment pressure with the nozzle discharge pressure. This operation modality is called underexpanded. The overexpanded modality produces stronger external turbulence than the underexpansion case.

The overexpanded and the underexpanded cases are also identified by a flow shape different to nozzle's shape. This will be discussed in section 5.3.

o This section is summarized as follows: shock waves produce a significant loss of kinetic energy plus a turbulent regime created by normal internal or oblique external shock waves. This is only mitigated by working with discharge pressures as close as possible to the design pressure.

4.7. Example of shock wave calculation

Refer to file 4.8. Tables 4.4 and 4.5. Shock wave calculation.xlsx.

In this example, the impact of a normal shock wave on the flow velocity and the kinetic energy is determined.

Requested information

o Determine whether the actual discharge pressure can produce normal shock waves.
o Calculate the flow properties at the shock wave outlet.
o Calculate the input and output speed of sound.
o Calculate the shock wave input and output velocities (V_y and V_z) with the Mach number formula $V = M.V_s$.
o Calculate the shock wave input and output kinetic energies (E_{ky} and E_{kz}) with $V^2/(2.g.J)$.
o Calculate the % loss of velocity and kinetic energy, before and after the shock wave.

Input data to calculate the shock wave

See Input Data in Table 4.4.

Calculation procedure of the shock wave

Seethis procedure implemented in Table 4.5.

o   First calculate M_z because output state variables are function of output Mach number M_z.
o   Second step is the input and output velocities calculation and % velocity loss across the shock wave.
o   Third step is the input and output energy calculation.

Table 4.4. Shock wave calculation. Input data

Input data		
Property name	**Input data**	**Units**
Discharge design pressure pD	2.00	kg/cm2 a
Discharge actual pressure pA	18.00	kg/cm2 a
Shock wave pressure at the nozzle outlet pB	13.29	kg/cm2 a
Shock wave location xw	4.60	cm
Shock wave input pressure py	2.61	kg/cm2 a
Shock wave input temperature Ty	262	°K
Mach number at the shock wave inlet My	2.25	NA

Conclusions of calculations made in Table 4.5

The discharge pressure of this example is abnormally high if compared with the design pressure, but this big difference was intentionally done to demonstrate the devastating effect of a shock wave on the flow discharge velocity and kinetic energy.

In this case the velocity loss is 65.4% and the kinetic energy loss is 88.1% at the shock wave outlet. However, as the divergent part of the nozzle after the shock wave works as a diffuser further losses of velocity and kinetic energy should be expected at the nozzle discharge. This means that at the outlet things will worsen. Surely a turbine can't work with at least 88.1% smaller energy in the incoming flow into the blades wheel.

Table 4.5. Shock wave calculation

Property	Formula	Results
Output Mach number	$M_z = \sqrt[2]{\left[\dfrac{2 + (k-1) \cdot M_y^2}{1 - k + 2 \cdot k \cdot M_y^2} \right]}$	0.541
Output pressure Kg/cm² a	$p_z = p_y \cdot \left[\dfrac{1 + k \cdot M_y^2}{1 + k \cdot M_z^2} \right]$	14.94
Output amplitude Kg/cm² a	$\text{Amplitude} = p_y - p_z$	12.32
Output temperature, °K	$T_z = T_y \cdot \left[\dfrac{2 + (k-1) \cdot M_y^2}{2 + (k-1) \cdot M_z^2} \right]$	538
Input sound velocity, m/s	$V_{sy} = \sqrt[2]{g \cdot k \cdot J \cdot R \cdot T_y}$	374
Input velocity, m/s	$V_y = M_y \cdot V_{sy}$	839
Output sound velocity, m/s	$V_{sz} = \sqrt[2]{g \cdot k \cdot J \cdot R \cdot T_z}$	536
Output velocity, m/s	$V_z = M_z \cdot V_{sz}$	290
Velocity % loss	$L_{V\%} = \dfrac{V_y - V_z}{V_y} \times 100$	-65.4
Input kinetic energy, kJ/kg	$E_{ky} = \dfrac{V_y^2}{2 \cdot g \cdot J}$	352
Output kinetic energy, kJ/kg	$E_{kz} = E_{ky} \cdot \left(\dfrac{V_z}{V_y} \right)^2$	42
Kinetic energy % loss	$L_{KE\%} = \dfrac{E_{ky} - E_{kz}}{E_{ky}} \times 100$	-88.1%

4.8. Example of a turbine power affected by a normal shock wave

Input data

The typical operation of the nozzle calculated in section 4.7 is as follows.

- The nozzle is installed in a gas turbine, which in total has 6 identical nozzles.
- Minimum expected net delivery power of the turbine is 850 kW
- Discharge pressure $\cong$ Design pressure $(p_2 \cong p_D)$
- Mass flow: $G = 0.5$ kg/s per nozzle
- Actual kinetic energy at the outlet: $E_{k2} = 359$ kJ/kg per nozzle
- Nozzle length: $L_t = 6.23$ cm
- Turbine efficiency: 90%

Determine

- Case a. The power delivered by the turbine with no shock waves in the nozzles
- Case b. The power delivered by the turbine with shock waves calculated in section 4.7.

Calculation

The turbines power formula is:

Formula 4.18. Turbine power

$$P = N \cdot G \cdot E_{k2} \cdot e_t \qquad \left| kW \right| or \left| \frac{kJ}{s} \right|$$

Where:

P: turbine shaft power
N: nozzles quantity
e_t: turbine efficiency

○ Case a.

$$P = 6 \times 0.5 \times 359 \times 0.90 = 969.8\text{kW}$$

This power is 14% higher than the expected power indicated in the Input Data of this example. Hence, the typical operating conditions are fulfilled.

○ Case b.

As the shock wave is close to the nozzle outlet, it's assumed that the shock wave energy at the nozzle outlet equals the energy at the shock wave outlet. Hence, the kinetic energy outlet is read in Table 4.5. This assumption is not conservative, because the kinetic energy will be additionally reduced by the diffuser effect, which happens between the shock wave outlet and the nozzle outlet. Anyway, the following calculation demonstrates that even in this not conservative assessment, the total power delivered by the six nozzles is very poor.

$$P = 6 \times 0.5 \times 42.0 \times 0.90 = 113.4\text{kW}$$

The conclusion of this result is that the turbine power has plummeted by approximately 88%. Usually this power reduction is completely unacceptable. A fast solution to this case is to make efforts to return to a discharge pressure value close to the design pressure.

Chapter 5

Geometric design and performance curves

5.1. Cross section area calculation

For this section refer to file 5.1. Tables 5.1 and 5.3. Area ratios tables.xlsx.

The basic formula to calculate the nozzle profile is derived from the principle of continuity. By clearing A from Formula 1.5, the resulting expression for any cross section area is as follows.

Formula 5.1. Area formula according to principle of continuity

$$A_x = G \cdot \frac{v_x}{V_x} \qquad \left| cm^2 \right|$$

This formula has two independent variables: flow velocity and specific volume. However, it's possible to reduce it to two different formulas with only one independent variable; the pressure ratio or the Mach number. Both formulas are explained in sections 5.1.1 and 5.1.3 respectively and can be used indistinctly, because they return the same result.

For convenience, the cross section area is usually calculated on the basis of the A_x/A_t ratio. Therefore, the throat area is used as the reference for the nozzle profile determination. From Formula 5.1, the areas ratio is given by:

Formula 5.2. Cross section areas ratio versus v and V

$$\frac{A_x}{A_t} = \frac{v_x}{v_t} \cdot \frac{V_t}{V_x}$$

Formula 5.2 is conveniently modified using Formula 3.3 and 1.9. The result is Formula 5.3, valid for gas and steam.

Formula 5.3. Cross section areas ratio versus
p, T and h. Valid for gas and steam

$$\frac{A_x}{A_t} = \frac{T_x}{T_t} \cdot \frac{p_t}{p_x} \cdot \sqrt[2]{\frac{h_1 - h_t}{h_1 - h_x}}$$

In any case, the throat area is calculated with Formula 5.1 and used as the reference for all other cross section by means of Formula 5.3. This formula is important because it links the thermodynamic properties and the nozzle geometry together. This means that the nozzle profile is determined on the basis of the thermodynamic state of the fluid.

5.1.1. Formula of area versus pressure ratio r_{px} for ideal gas

Formula 5.3 can be converted into a function of only one independent variable. However, the resulting formula is only valid for gas and not for steam. The derivation of this formula can be consulted in Chapter 8. The result is the following expression.

Formula 5.4. Cross section areas ratio
versus r_{px}. Only valid for gas

$$\frac{A_x}{A_t} = \left(\frac{r_{pt}}{r_{px}}\right)^{\frac{1}{k}} \cdot \sqrt[2]{\frac{1 - \left(r_{pt}\right)^{\left(1-\frac{1}{k}\right)}}{1 - \left(r_{px}\right)^{\left(1-\frac{1}{k}\right)}}}$$

Where:

$r_{pt} = \dfrac{p_t}{p_1}$: Pressure ratio at the throat. In CD nozzles this ratio is the critical ratio given by Formula 3.7. Critical pressure ratio

$r_{px} = \dfrac{p_x}{p_1}$: Pressure ratio at any location x.

The main property of this curve is its U-shaped appearance, which resembles the CD nozzle profile. The extreme branches go asymptotically to infinity, but these extreme branches have no technical importance because they are produced by no realistic assumptions, which would be zero output pressure or outlet pressure equal to input pressure (right branch). None of these situations have practical existence. Actually, the nozzle design requires to be specified. In this section's example the chosen r_{po} range is from 0.20 to 0.95. This means that the nozzle to be designed will never work at a discharge pressure lower than 20% of the input pressure nor more than 95% of the same pressure.

Figure 5.1 indicates that as nozzle ratio is lower, cross sections area of divergent part are bigger. The minimum of the curve is produced at the critical pressure. Therefore, that minimum is the areas ratio for the throat section.

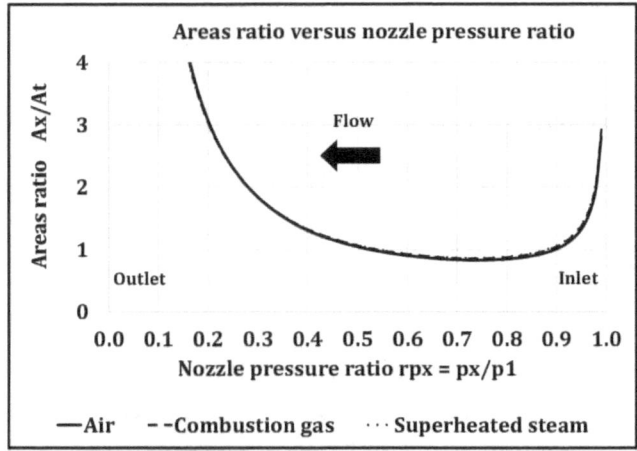

Figure 5.1. Areas ratio versus pressure ratio

Figure 5.1 demonstrates that the fluid type does not affect the area calculation versus the nozzle pressure ratio. Figure 5.1 includes superheated steam. As was said before, superheated steam could be considered an ideal gas, hence, Formula 5.4 is used to calculate cross section areas of steam nozzle. The result is an acceptable approximation.

5.1.2. Example of nozzle profile and cross sections area calculation

The following example is an application of Formula 5.4.

This calculation is in file 5.2. Figure 5.2. CD nozzle profile versus rp.xlsx. The input data is in Table 5.1 and the calculation procedure in Table 5.2.

With the data of Table 5.1 determine the following specifications of a conical nozzle.

- Throat area A_t
- Curve of sections area versus pressure ratio r_{po}

o   Nozzle profile for the discharge pressure range (20 to 95% of p_2)

Table 5.1. Input data and physical constants

Input data			Physical constants		
Mass flow G	1.0	kg/s	J	102	kg.m/kJ
Input temperature	1,300	°K	k air	1.40	
Input pressure	9.50	kg/cm2 a	cp at 1,100°K	1.155	kJ/|kg.°K|
Discharge pressure range			g	9.81	m/s2
Minimum % of p1	20%				
Maximum % of p1	95%				
Minimum p2	1.900	kg/cm2 a			
Maximum p2	9.025	kg/cm2 a			

Table 5.2. Throat area calculation

Property	Formula	Result
Specific constant of air kJ/|kg.°K|	$R = c_p \cdot \left(1 - \dfrac{1}{k}\right)$	0.330
Throat pressure ratio	$r_{pc} = \dfrac{p_c}{p_1} = \left(\dfrac{2}{k+1}\right)^{\frac{k}{k-1}}$	0.528
Critical pressure kg/cm² a	$p_c = r_{pc} \cdot p_1$	5.02
Nozzle type	$=IF(p_{2\,min} < p_{c,}$ "CD type", "C type")	CD type
Throat temperature ratio	$r_{tc} = \dfrac{T_c}{T_1} = \dfrac{2}{k+1}$	0.833
Throat temperature °K	$T_t = r_{tc} \cdot T_1$	1,083

Throat specific volume m³/kg	$V_t = \dfrac{J \cdot R \cdot T_t}{\left(10^4 \cdot p_c\right)}$	0.727
Velocity at the throat, m/s	$V_t = \sqrt[2]{2 \cdot g \cdot J \cdot c_p \left(T_1 - T_t\right)}$	708
Speed of sound at the throat, m/s	$V_s = \sqrt[2]{g \cdot k \cdot J \cdot R \cdot T_t}$	708
Throat area, cm²	$A_t = G \cdot \dfrac{V_t}{V_t} \cdot 10^4$	10.27

As now A_t is known after the calculation of Table 5.2, Formula 5.4 is used to plot the nozzle profile versus pressure ratio. Figure 5.2 represents the nozzle profile and the sections area curve. This curve resembles same shape as the curves shown in Figure 5.1. However, coordinates of the former is the A_x/A_t ratio and the coordinates of Figure 5.2 is the A_x area curve.

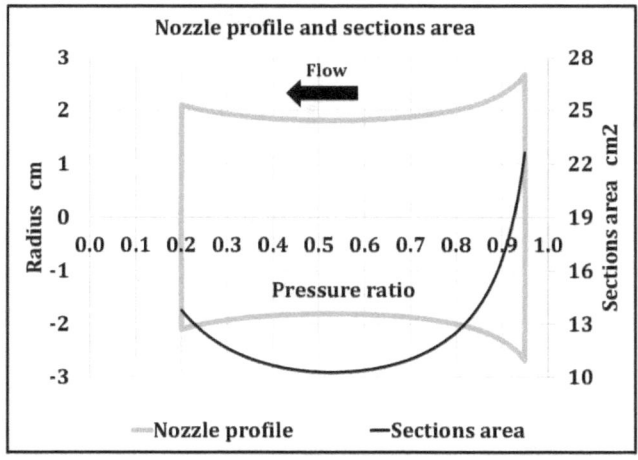

Figure 5.2. Nozzle profile and sections area
curve versus pressure ratio

The conversion of this profile to a graph versus abscissa x, is explained in section 5.7.3. That representation will show the actual geometric profile of the nozzle.

5.1.3. Formula of area versus M_x

The Mach number is defined as the ratio between the fluid velocity in a vein section and the speed of sound in the same vein section. The Mach number formula is obtained by dividing flow velocity over speed of sound.

Formula 5.5. Mach number

$$M_x = \frac{V_x}{V_{sx}} = \frac{\sqrt{2 \cdot g \cdot J \cdot (h_1 - h_x)}}{\sqrt[2]{k \cdot g \cdot J \cdot R \cdot T_x}}$$

Formula 5.5 leads to the following formula, which is derived in section 8.8. This formula represents the A_x/A_t ratio versus the Mach number at section x.

Formula 5.6. Areas ratio A_x/A_t versus Mach number

$$\frac{A_x}{A_t} = \left[\frac{2 + (k-1) \cdot M_x^2}{2 + (k-1) \cdot M_t^2} \right]^{\frac{k+1}{2 \cdot (k-1)}} \cdot \frac{M_t}{M_x}$$

As a CD nozzle always operates under choked flow condition, the application of Formula 5.6 is simplified by doing $M_t = 1$.

M_x distribution along the x axis is calculated with the flow velocity and speed of sound distributions. This distribution will be explained in sections 5.6 and 5.7 and also in Chapter 6 and Chapter 7. The throat area is calculated according to the procedure of Table 5.2. Once the throat area A_t and a table of M_x

distribution are known, the A_x sections area versus the Mach number are calculated with Formula 5.6.

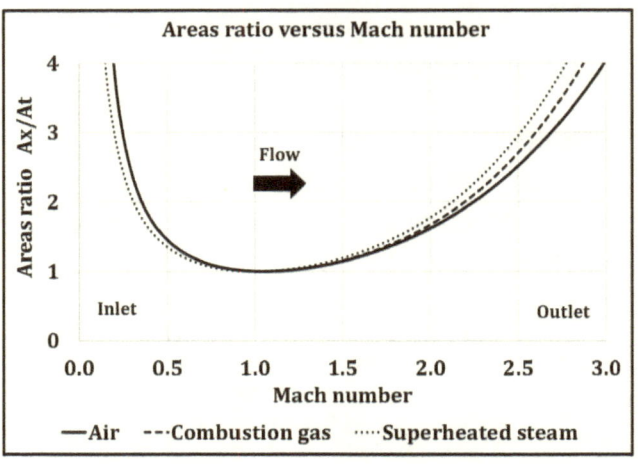

Figure 5.3. Areas ratio versus Mach number

Section areas of the divergent part are bigger as the k value of the fluid is lower. In other words, an ideal gas requires a smaller nozzle than any other gas, with lower k than the air. Furthermore, the size difference happens in the divergent part of the nozzle, while convergent part and throat dimensions with actual gas are practically equal to dimensions with ideal gas.

5.1.4. Formula of inlet area for CD conical nozzles

Formula 5.1, derived from the principle of continuity, is not applicable to inlet area calculation because usually input velocity is zero and the resulting area would be infinite. However, Formula 4.7 allows to calculate inlet area on the basis of the following considerations.

By clearing the L/D ratio of Formula 4.7 the following expression is obtained.

Formula 5.7. L/D ratio formula

$$\frac{L}{D} = \frac{1-\varphi^2}{f}$$

Where D is equal to $D_{avg} = \left(D_1 + D_t\right)\Big/2$.

The divergent part length is calculated with the following formula presented in Chapter 6.

Formula 5.8. Convergent part length

$$L_1 = \frac{D_1 - D_2}{2 \cdot tg\alpha_1} \qquad |cm|$$

Operating Formulas 5.7 and 5-8 to clear the final expression of D_1, the result is as follows.

Formula 5.9. Inlet diameter for a CD conical nozzle

$$D_1 = \frac{f + \tan\alpha_1 \cdot \left(1 - \varphi^2\right)}{f - \tan\alpha_1 \cdot \left(1 - \varphi^2\right)} \cdot D_t \qquad |cm|$$

Where:

o   f is the friction coefficient given by the Moody diagram
o α_1 is prefixed in the design stage, usually between 5° and 15°
o φ is usually specified also in the design stage

This procedure does some assumptions, which might not be entirely valid. These assumptions are: diameter D of Formula 5.7 is equivalent to the average between D_1 and D_t and friction factor is calculated with the Moody diagram or Moody formula, which were not developed for variable cross section pipe.

Anyway results of Formula 5.9 can be considered an acceptable approximation, especially in high velocity coefficients. Another caveat is that the results of Formula 5.9 are usually higher than other procedures.

The following example will help to understand the application of Formula 5.9.

Calculate the inlet area of a CD conical nozzle according to the following specifications:

o   Number of Reynolds: 80,000
o   Roughness of inner walls: 0.02
o   Converging angle: 40°
o   Velocity coefficient φ = 0.985
o   Throat diameter, D_t = 2 cm

The calculation procedure is as follows:

o   1st step. From Moody diagram: friction factor f = 0.05
o   2nd step. Use Formula 5.9:

$$D_1 = \frac{0.05 + \tan 40° \times \left(1 - 0.985^2\right)}{0.05 - \tan 40° \times \left(1 - 0.985^2\right)} \cdot 2 = 5.994 \text{cm}$$

o   3rd step. Inlet area calculation: A_2 = $\pi \times 5.994^2/4$ = 28.22 cm²

5.2. Flow shape and optimum design condition

The main condition of an optimum designed nozzle is that the nozzle geometric shapes and dimensions accommodate the flow with minimum turbulence. This condition is achieved when the nozzle and flow shapes and dimensions are the same. This configuration minimizes turbulences produced by the boundary layer separation, just as a wing profile reduces the aerodynamic resistance. As turbulence adversely impacts the velocity coefficient, the nozzle geometric design must prevent, as much as possible, turbulence formation. Figure 5.4 (which is a repetition of Figure 1.9) shows schematically the flow shape outgoing from an orifice, such as a nozzle throat. The flow straight section has initially a descendant trend until it reaches a minimum size at the throat. After the throat, the flow continues shrinking until a minimum section is produced. This minimum section is known as vena contracta.

Flow sections after the vena contracta begin to expand; it's the divergent part of the nozzle. If the nozzle shape and dimensions do not exactly match the flow shape and dimensions, as the case shown in Figure 5.4, turbulence is formed in the fluid layers close to the nozzle inner wall. Then, the fluid layers detach from the nozzle wall and this detachment inexorably creates turbulence zones, where the boundary layers become thicker. Actually, turbulence is also formed if the nozzle and flow shapes fit each other, but in this case turbulence is minimal.

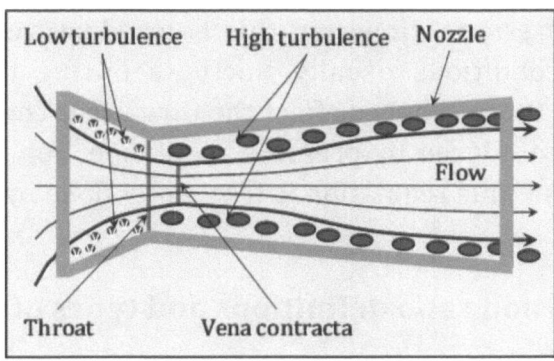

Figure 5.4. Flow shape and turbulence inside a conical flow.

(Repetition of Figure 1.9)

The flow shape and area depend on their thermodynamic and kinematic conditions (specific volume, mass flow and velocity) and are usually calculated with a formula derived from Formula 1.5.

Formula 5.10. Cross section area of flow

$$A_f = \frac{G \cdot v}{V} \qquad \left| m^2 \right|$$

If the nozzle operation is according to the design conditions this formula gives the optimum geometric profile.

The aftermath of the above considerations is that the optimum design condition is achieved when the flow area A_f matches the geometric nozzle area A_g, at any internal section.

Formula 5.11. Optimum design condition

$$A_f(x) = A_g(x) \qquad \left| m^2 \right|$$

This condition is achieved only by nozzles operating at the design discharge pressure. However, this is an ideal case, because operating conditions usually fluctuate during the turbine service and those changes affect the flow area. These changes are usually produced by pressure regulation. For instance, in steam turbine this regulation is frequently done by regulating the neck valve.

5.3. Expansion ratio definitions and types of expansion

Refer to file 5.4. Figure 5.5. Expansion identification.xlsx.

Expansion ratios indicate how big the outlet area is respect to the throat area of a nozzle or its outgoing flow.

The geometric expansion ratio X_g equals the outlet cross section area over the throat cross section area.

Formula 5.12. Geometric expansion ratio

$$X_g = \frac{A_2}{A_t}$$

This ratio is also known as actual expansion ratio. If the nozzle is already manufactured, this ratio is a constant in the nozzle assessment formulas. In the design phase, this value is set as appropriate. Flow expansion ratio or theoretical expansion are equivalent denominations. It's the outlet flow area given by Formula 5.1 over the throat flow area. According to Formula 5.2 flow expansion is calculated with the following formula.

Formula 5.13. Flow expansion ratio

$$X_f = \frac{A_{2f}}{A_{tf}} = \frac{v_2}{v_t} \cdot \frac{V_t}{V_2}$$

If the optimum design condition of Formula 5.11 is satisfied, it follows that the flow and geometric expansion ratios must be equal.

Formula 5.14. Optimum relation between geometric and flow expansions

$$X_f = X_g$$

If the optimum relation of Formula 5.14 is achieved, nozzle tests demonstrate that the velocity coefficient φ is maximal.

From the above definition of expansion ratios, it follows that there exist three types of expansion curves, which are plotted in Figure 5.5.

5.3.1. Exact expansion: $X_g = X_f$

Exact expansion means that the actual discharge pressure coincides with the design pressure.

In this example the flow has little turbulence because its shape and dimensions match the nozzle shape and dimensions. This means that the optimum condition of Formula 5.11 is fully achieved. This result, of course, must be supported by a good physical nozzle integrity and a high quality polishing of the nozzle inner surface to assure the nozzle efficiency.

5.3.2. Overexpansion. $X_g > X_f$

In this case the nozzle dimensions are higher than the theoretical flow dimensions calculated with Formula 5.1. However, "higher nozzle dimensions" doesn't mean that there are empty spaces between the theoretical flow and the nozzle internal walls. Actually the fluid expands completely filling the nozzle with turbulent fluid veins. In this case, sudden rise of pressure takes place inside the nozzle at the expense of the kinetic energy of the flow with a strong turbulent flow.

5.3.3. Underexpansion. $X_g < X_f$

In this case the nozzle dimensions are smaller than the theoretical flow dimensions. This dimension's difference occurs when the actual discharge pressure is smaller than the design pressure p_D, therefore, the design pressure is reached inside the nozzle and before the outlet section. This means that the flow is underexpanded.

The aftermath of this underexpansion is that the internal flow is composed of turbulent fluid veins, although this turbulence isn't as severe as in the overexpansion case. At the outlet section a sudden pressure drop occurs and shock waves are formed, not within the nozzle but outside it. As explained before, these external shock waves are undulating and oblique, and not normal to the flow direction.

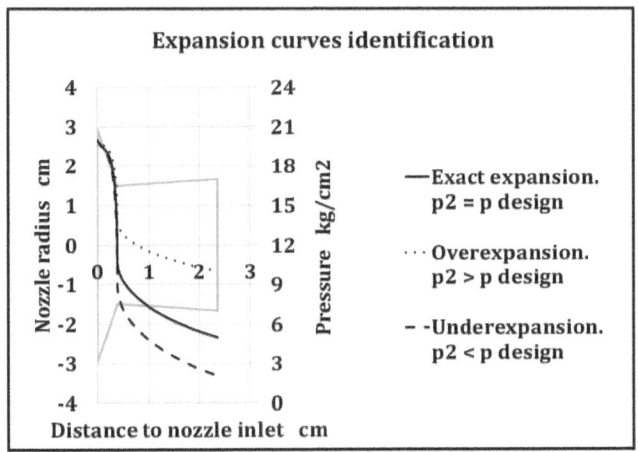

Figure 5.5. Identification curves of nozzle expansion

This situation occurs because the actual discharge pressure is between points B and D of Figure 4.7 and; consequently, an oblique shock wave is formed outside the nozzle. Therefore, the design pressure does not fall within the nozzle but outside it, hence, further expansion outside the nozzle is necessary to reach the design pressure.

The term "underexpansion" is used after this situation, where additional expansion happens outside the nozzle.

5.3.4. Notes to expansion types

As turbulence created by overexpansion is more severe than underexpansion turbulence, under high demand conditions it is advisable to operate with the underexpansion regime and in low demand conditions with the overexpansion regime. This modality minimizes losses because the discharge pressure is located in a range around the design pressure p_D.

The efficiency loss due to turbulence, has discouraged the use of CD nozzles in steam turbines that operate with variable loads. Nonetheless CD nozzles are used in the regulating stages of multistage turbines. Thus an important advantage of the CD nozzles is used, which is the supersonic flow velocity released by them. As was seen before, this high velocity does not happen in converging nozzles.

5.4. Velocity coefficient. US Naval Institute

For this section refer to file 5.5. Figures 5.6 to 5.9. Velocity coefficient and efficiency charts.xlsx.

Numerous tests in steam CD nozzles have demonstrated that there exists a mathematical relationship between the velocity coefficient and the flow and geometric expansion ratios. Consequently, as velocity coefficient depends on these ratios, nozzle efficiency is also dependent on them, as stated by Formula 2.9. Therefore, the expansion ratios are used to calculate nozzle efficiency and to optimize nozzle geometric design.

The procedure presented in the Energy Analysis of Naval Machinery book, published by the US Naval Institute, establishes in Figure 17-7 that the velocity coefficient in a CD nozzle is a function of the two expansion ratios X_f and X_g. Although formula of φ is not disclosed in the above mentioned book, Figure 17-7 shows a chart composed of 16 curves in double logarithmic scale. Each curve identifies a velocity coefficient value and is intended to calculate velocity coefficient. This chart is reproduced in Figure 5.6.

The horizontal axis represents the geometric expansion ratio X_g. The ordinate represents the flow expansion ratio. The graph is prepared for nozzles with 94% maximum efficiency or maximum velocity coefficient of 0.97. Nonetheless, similar charts are produced for the same nozzle type, whose maximum velocity coefficient is higher, or lower, than 0.97. The velocity coefficient is picked off in the intersection of X_g and X_f coordinates. For example, for $X_g = 4$ and $X_f = 6$, the resulting velocity coefficient is 0.92.

For this book, chart of Figure 5.6 has been mathematically processed to obtain a simpler chart. The result of this mathematical procedure is one formula of velocity coefficient that only depends from the X_g/X_f ratio. See Formula 5.15. Therefore, velocity coefficient is represented by only one curve, as shown in Figure 5.7.

Formula 5.15. Velocity coefficient for nozzle steam

$$\varphi = 9.934 \cdot r_A^6 - 61.401 \cdot r_A^5 + 160.388 \cdot r_A^4 - 214.645 \cdot r_A^3 + 159.389 \cdot r_A^2 - 60.999 \cdot r_A + 10.304$$

Where r_A is the expansions X_g/X_f ratio or the outlet areas A_{2g}/A_{2f} ratio. These properties are linked by the following formula:

Formula 5.16. Expansions ratio or outlet areas ratio

$$r_A = \frac{X_g}{X_f} = \frac{A_{2g}}{A_{2f}} \cdot \frac{A_{tf}}{A_{tg}} = \frac{A_{2g}}{A_{2f}}$$

According to Formula 5.13 the expansions ratio X_g/X_f can be expressed in terms of outlet areas ratio only. This is possible because the flow area and the geometric area are equals at the throat. This equality exists because the flow is choked. Hence, regardless of changes at the outlet, no changes happen at the throat and consequently, $A_{tf} = A_{tg}$ always.

However, this procedure has an issue. As was said before the overexpansion condition produces a higher turbulence, which means that velocity coefficient values are lower than in the underexpansion regime. This behavior is not apparent in the curve of Figure 5.7, because it predicts, for the overexpanded condition velocity, coefficients of the same order than the underexpanded area. The aftermath is that the author does not recommend to use the curve of Figure 5.6 and 5.7. In the next section the Steinmetz curve will be discussed, which demonstrates, better than the above mentioned figures, the disadvantages of overexpansion over underexpansion regimes.

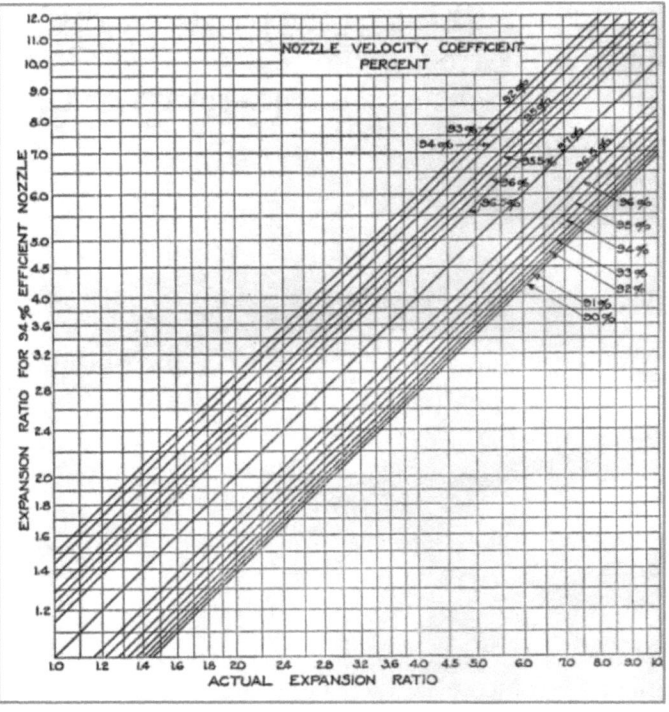

Figure 5.6. Velocity coefficient. Steam
nozzles. US Naval Institute chart

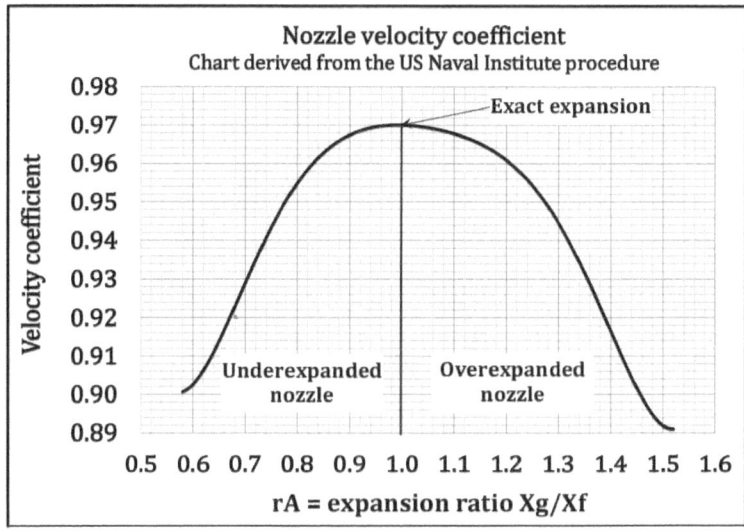

Figure 5.7. Nozzle velocity coefficient for steam nozzles.

5.5. Velocity coefficient. C. P. Steinmetz curve

For this section refer to file 5.5. Figures 5.6 to 5.9. Velocity coefficient and efficiency charts.xlsx.

A paper by C. P. Steinmetz presents two curves, one of velocity coefficient reduction and the other of energy % losses versus the A_g/A_f ratio. The Steinmetz paper states that these curves correspond to average practice conditions and φ_{max} equal to unity, which could be considered optimistic. Refer to book Steam Turbines by Edwin F. Church, Figure 82. From this figure a chart of velocity coefficient versus areas ratio A_g/A_f was derived and the result is presented in Figure 5.8.

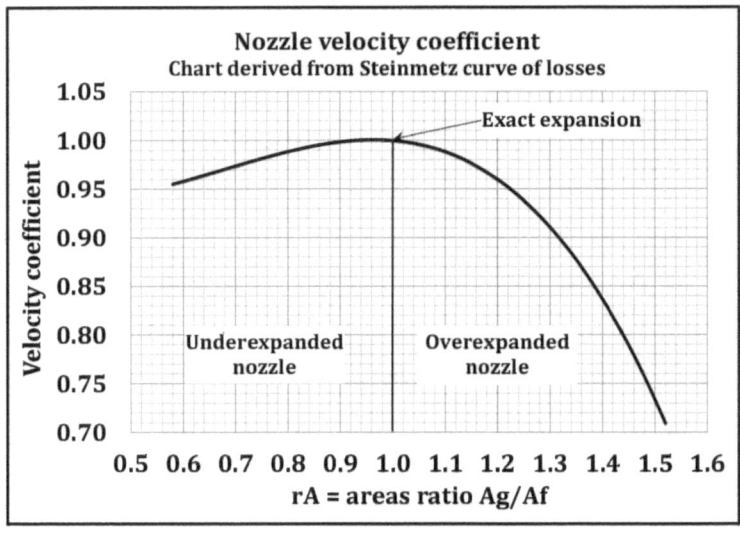

Figure 5.8. Steinmetz curve

The velocity coefficient formula derived
from the Steinmetz curve is:

Formula 5.17. Velocity coefficient formula for Steinmetz curve

$$\varphi = -0.669 \times r_A^{\,3} + 1.367 \times r_A^{\,2} - 0.771 \times r_A + 1.073$$

Where r_A is also the outlet areas ratio as defined in Formula 5.16. The curve of Figure 5.8 clearly shows that an overexpanded nozzle has lower velocity coefficients than underexpanded nozzles. In Figure 5.9 are represented the US Naval Institute curve and the Steinmetz curve, only for comparison purposes. The last is the same than Figure 5.8, but downwardly displaced until both curves have the same maximal value, that is φ_{max} = 0.97. Although this is not an exact procedure, the resulting graph is an acceptable comparison between both the USA Naval Institute and the Steinmetz curves.

As Steinmetz curve is less conservative in the overexpanded region, the author recommends to use it as a first approach to a velocity coefficient calculation, which is applicable to steam nozzles with φ_{max} = 0.97. See Figure 5.9.

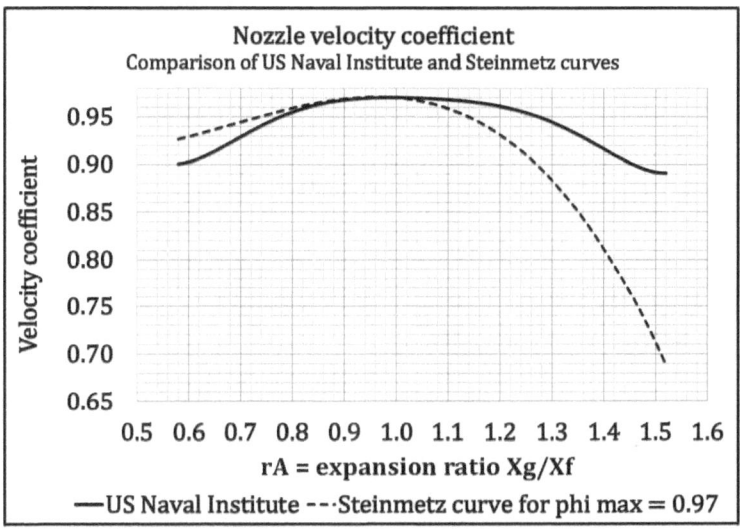

Figure 5.9. Steinmetz and US Naval Institute curves with same maximum velocity coefficient

Figure 5.9 suggests the following conclusions:

○   The US Naval Institute curve is symmetric and almost flat. As was said before, it means that velocity coefficient has no sensibility to the difference between underexpanded and overexpanded regimes. This is not consistent with the turbulence phenomena inside a nozzle, which is produced by discharge pressures different to design pressure.

○   Both curves present a kind of similarity for underexpanded regime. Although the Steinmetz curve is higher than the US Naval Institute curve, this difference is not that much. In both cases velocity coefficient is higher than 0.90.

○   The Steinmetz curve suddenly plunges in the overexpanded zone, just as predicted by the turbulence study.

○ This comparison is supposed to be valid for the same steam CD nozzle. Anyway, both curves are helpful to understand the velocity coefficient variations versus the expansion regime for any other type of CD nozzle.

○ The aftermath is that the Steinmetz curve appears to be more realistic than the US Naval Institute curve. Therefore, the Steinmetz curve must be preferred for practical applications, provided that further reliable information is not available.

5.6. Geometric design procedure

The first step in nozzle design is to decide the nozzle type. This type is strongly dependent on the turbine requirements, such as flow velocity, mass flow, etc.

To decide the nozzle type it's mandatory to calculate the critical pressure at the throat. After this property is known it must be compared to the discharge pressure required by the turbine. It was explained before that for critical pressures higher than the design discharge pressure, the nozzle type must be CD. Otherwise a C type nozzle must be used.

Once the nozzle type (C or CD) has been defined, a simple calculation procedure is implemented in a spreadsheet to design a nozzle profile to match the flow shape. In other words, thermodynamic calculations provide the shape of the flow and the nozzle design intends to sheathe that shape with an identical metallic and sturdy coverage. That sheath is the nozzle.

The calculation flow chart is represented in Figure 5.10. It shows four calculating blocks, being the last one the geometric dimensions' calculator. This geometric calculation is done on the basis of the shape of the flow given by Blocks B and C.

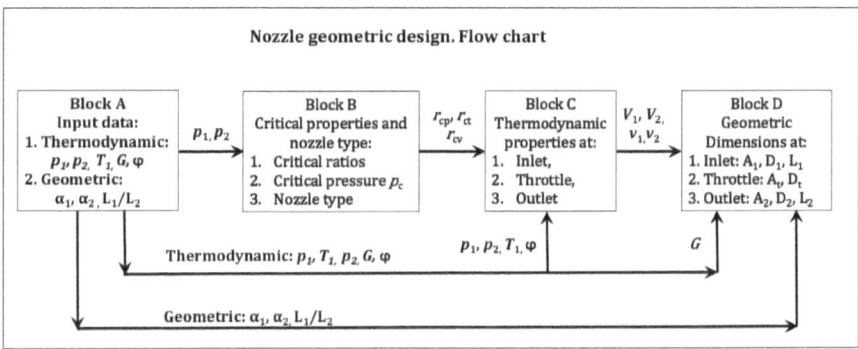

Figure 5.10. Nozzle design flow chart

The applicable formulas for each block are included in the example of this section. These formulas are used later to create the performance curves in a spreadsheet.

5.7. Example of a CD nozzle design

For this example, refer to files:

5.7. Figure 5.11. Nozzle dimensions sensitivity.xlsx

5.8. Figure 5.11 to 5.15. Example of a CD nozzle geometric design. xlsx

The applicable formulas and results of this design are summarized in Figure 5.10 and in Tables 5-4, 5-5, 5-6 and 5-7.

5.7.1. Technical specifications

On the basis of data of Table 5.3 determine:

○ Critical ratios and nozzle type identification:
○ Thermodynamic properties at the inlet, throat and outlet, under ideal and actual conditions.

o   Design the geometric dimensions of two different nozzles: one under ideal conditions ($\varphi = 1$) and the other under actual conditions ($\varphi = 0.90$).
o   Performance curves: p_x, v_x, T_x, h_x, V_x. Draw the nozzle profile along with the pressure curve
o   Calculate the expansion ratios under ideal and actual conditions.
o   Produce a graph of the nozzle volume and length versus the velocity coefficient.

Use: φ = 0.90, 0.92, 0.94, 0.96, 0.98 and 1.00

Note: The L_1/L_2 ratio, specified to calculate the inlet area, is not a universal recommended value. It's valid only for this case.

Table 5.3. Input data and physical constants

Thermodynamic data		Physical constants	
Input pressure p_1 kg/cm2 a	20.0	Specific heat c_p kJ/|kg.°K|	1.248
Discharge pressure p_2. kg/cm2 a	5.0	Specific heats ratio	1.330
Input temperature T_1. °K	1,200	Specific gas constants R kJ/|kg.°K|	0.3095
Mass flow G kg/s	1,50	Mechanical equivalent of heat J kg.m/kJ	102.0
Velocity coefficient φ	0.90		
Convergent angle α_1	75°		
Divergent angle α_2	5°		
Lengths ratio L_1/L_2	0.20		

Table 5.4. Critical ratios, critical pressure and nozzle type identification		
Property	**Formula**	**Result**
Pressure	$r_{pc} = \dfrac{p_c}{p_1} = \left(\dfrac{2}{k+1}\right)^{\frac{k}{k-1}}$	0.5404
Temperature	$r_{tc} = \dfrac{T_c}{T_1} = \dfrac{2}{k+1}$	0.8584
Specific volume	$r_{vc} = \dfrac{v_c}{v_1} = \left(\dfrac{k+1}{2}\right)^{\frac{1}{k-1}}$	1.5885
Nozzle pressure ratio	$r_{pn} = \dfrac{p_2}{p_1}$	0.2500
Critical pressure	$p_c = r_{pc} \cdot p_1$	10.81 kg/cm2 a
Nozzle type identification: As $p_2 < p_c$, the flow is choked flow, then nozzle type is CD		

As was said before, the principle of continuity can't be used because the inlet velocity equals zero and that would produce an infinite inlet area according to Formula 1.5. There are other procedures to work out this issue and one of them is presented in this example, which is to prefix the L_1/L_2 ratio.

Table 5.5. Thermodynamic properties

Property	Inlet	Throat	Outlet Ideal values
Pressure kg/cm2 a	$p_1 = 20.0$ From Table 5.3	$p_c = 10.81$ From Table 5-4	$p_2 = 5.0$ From Table 5-3
Temperature °K	$T_1 = 1{,}200$ From Table 5-3	$T_c = r_{tc} \cdot T_1 = 1{,}030$	$T_{2ideal} = T_1 \cdot \left(\dfrac{p_2}{p_1}\right)^{\frac{k-1}{k}}$ $= 851$
Specific volume m³/kg	$v_1 = \dfrac{J \cdot R \cdot T_1}{p_1 \cdot 10^4}$ $= 0.189$	$v_c = r_{tc} \cdot v_1$ $=0.301$	v_{2ideal} $= v_1 \cdot \left(\dfrac{p_1}{p_2}\right)^{\frac{1}{k}}$ $= 0.537$
Enthalpy. Ideal kJ/(kg.°K)	$h_1 = c_p \cdot T_1$ $= 1{,}497$	$h_t = c_p \cdot T_t$ $= 1{,}285$	h_{2ideal} $= c_p \cdot T_{2ideal}$ $= 1{,}061$
Velocity m/s	$V_1 = 0$ $V_t = \sqrt[2]{2 \cdot g \cdot J \cdot \left(h_1 - h_{t)}\right)} = 651$ $V_{2ideal} = \sqrt[2]{2 \cdot g \cdot J \cdot \left(h_1 - h_{2ideal}\right)} = 934$		
Velocity coefficient and efficiency	$\varphi = 0.9$ From Table 5.3	$e_n = \varphi^2 = 81.00$ %	
Total available energy. kJ/ (kg.°K)	$E_a = h_1 - h_{2ideal} = 436$		

Friction losses kJ/(kg.°K)	$q_f = (1 - e_n) \cdot E_a = 83$
Temperature increment produced by friction losses°K	$\Delta T_2 = \dfrac{q_f}{c_p} = 66$

Outlet actual values			
Temperature °K	Specific volume m3/kg	Enthalpy k/(kg.°K)	Velocity m/s
$T_2 = T_{2ideal} + \cdot T_2$ $=$ 917	$V_2 = \dfrac{J \cdot R \cdot T_2}{p_2 \cdot 10^4}$ $= 0.579$	$h_2 = c_p \cdot T_2$ $= 1,144$	V_2 $= \varphi \cdot V_{1ideal}$ $= 840$

This example demonstrates the reheat of the outgoing flow and its impact on the outlet properties. This impact is especially noticeable in the geometric design because as outlet velocity is smaller than the ideal case the outlet area must be higher, according to the principle of mass continuity. To bridge this issue some geometric characteristics of the nozzle converging portion have been prefixed in Table 5.7. In this example the prefixed properties are: α_1, α_2 and L_1/L_2 ratio. The prefixed values depend on the designer's experience and the practical information he has at hand.

The friction impact is on the specific volume and velocity at the outlet, which make the flow area to be larger than the ideal case. Consequently, the divergent part length is also larger than the ideal case according to the following trigonometric formula:

Formula 5.18. Divergent part length

$$L_2 = \frac{D_2 - D_t}{2 \cdot \tan(\alpha_2)}$$

Table 5.6. Nozzle 1. Geometric design under ideal conditions

Property	Formula	Value
Area. Throat cm²	$A_t = G \cdot \dfrac{v_t}{V_t} \cdot 10^4$	6.93
Diameter. Throat cm	$D_t = \sqrt[2]{4 \cdot A_t / ð}$	2.97
Area. Outlet Cm	$A_{2ideal} = G \cdot \dfrac{v_{2ideal}}{V_{2ideal}} \cdot 10^4$	8.63
Diameter. Outlet Cm	$D_{2ideal} = \sqrt[2]{4 \cdot A_{2ideal} / ð}$	3.31
Length. Divergent part. cm	$L_{2ideal} = \dfrac{D_{2ideal} - D_t}{2 \cdot \tan(\alpha_2)}$	1.97
Length. Convergent part cm	$r_L = 0.2$. From input data $L_{1ideal} = r_L \cdot L_{2ideal}$	0.39
Length. Total cm	$L_{t\,ideal} = L_{1ideal} + L_{2ideal}$	2.36
Inlet section diameter. cm	$D_1 = D_t + 2 \cdot L_c \cdot \tan(\alpha_1)$	5.91
Inlet section area cm²	$A_1 = \dfrac{\pi \cdot D_1^2}{4}$	27.42
Volume. Convergent part cm³	$Vol_{1ideal} = \dfrac{\pi \cdot L_1}{12} \cdot \left(D_1^2 + D_t^2 + D_1 \cdot D_t\right)$	6.32

Volume. Divergent part cm^3	$$Vol_{2ideal} = \frac{\pi \cdot L_2}{12}$$ $$\cdot \left(D_{2ideal}^2 + D_t^2 + D_{2ideal} \cdot D_t \right)$$	15.28
Volume. Total cm^3	$$Vol_{total\,ideal} = Vol_{1ideal} + Vol_{2ideal}$$	21.60

The geometric design under ideal conditions of Table 5.6 must be refined by adding the impact of the inner friction on the nozzle dimensions. The results of this calculation are shown in Table 5.7.

Table 5.7. Geometric design under actual conditions. φ = 0.9

Property	Formula	Result	% Variation
Area. Throat cm^2	$$A_t = A_{t\,ideal} = G \cdot \frac{v_t}{V_t} \cdot 10^4$$	6.93	0.0%
Diameter. Throat cm	$$D_t = D_{t\,ideal} = \sqrt[2]{\frac{4 \cdot A_t}{ð}}$$	2.97	0.0%
Area. Outlet cm^2	$$A_2 = G \cdot \frac{v_2}{V_2} \cdot 10^4$$	10.34	+19.8%
Diameter. Outlet cm	$$D_2 = \sqrt[2]{\frac{4 \cdot A_2}{\pi}}$$	3.63	+9.7%

Table 5.7. Geometric design under
actual conditions. $\varphi = 0.9$

Property	Formula	Result	% Variation
Length. Divergent part cm	$L_2 = \dfrac{D_2 - D_t}{2 \cdot \tan(\alpha_2)}$	3.76	+90.9%
Length. Convergent part cm	$r_L = 0.2.$ From input data $L_1 = r_L \cdot L_2$	0.75	+92.3%
Length. Total nozzle cm	$L_t = L_1 + L_2$	4.51	+91.1%
Inlet section diameter cm	$D_1 = D_t + 2 \cdot L_c \cdot \tan(\alpha_1)$	8.58	+45.2%
Inlet section area cm^2	$A_1 = \dfrac{\pi \cdot D_1^2}{4}$	57.81	+110.8%
Volume. Convergent part cm^3	$Vol_1 = \dfrac{\pi \cdot L_1}{12} \cdot \left(D_1^2 + D_t^2 + D_1 \cdot D_t\right)$	21.23	+235.9%
Volume. Divergent part cm^3	$Vol_2 = \dfrac{\pi \cdot L_2}{12} \cdot \left(D_2^2 + D_t^2 + D_2 \cdot D_t\right)$	32.23	+110.9%
Volume. Total nozzle cm^3	$Vol_{total} = Vol_1 + Vol_2$	53.64	+148.3%

The column named "% Variation" in Table 5.7, shows the nozzle dimensions respect to the ideal case. Table 5.7 shows a significant additional volume (+148.3%), which suggests a heavier nozzle and of course, also an extra cost per nozzle. An important aftermath is that the velocity coefficient drives not only the nozzle efficiency, but also the nozzle cost. Small variations in φ produce significant increments in the nozzle dimensions and consequently, in its materials weight and cost.

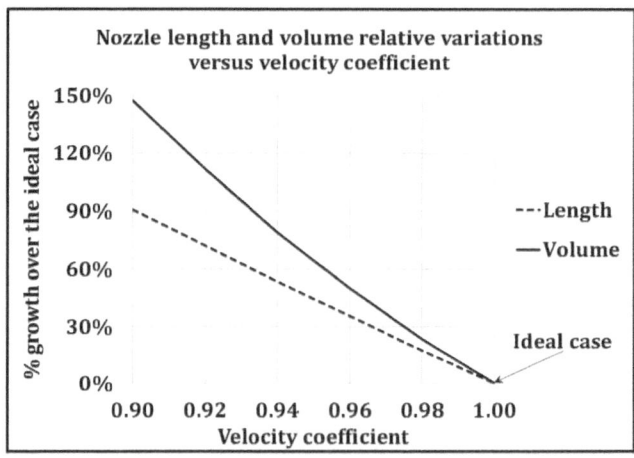

Figure 5.11. Example of conical nozzle dimensions versus velocity coefficient

See file 5.7. Figure 5.11. Nozzle dimensions sensitivity.xlsx. It was used to calculate nozzlelength and volume % growth versus the velocity coefficient shown in Figure 5.11.

For example, if the velocity coefficient changes from 1.00 to 0.96 the nozzle volume increases by 50% and the length growths by 36%. Gas nozzles usually have a velocity coefficient pretty close to 0.98. These 0.02 points less of the velocity coefficient over the ideal case, makes the nozzle to be 18% longer and have a volume 24% higher.

5.7.2. Expansion ratios

There is one important difference between an ideal design and a design under actual conditions, which is the nozzle dimensions. As was said before, a velocity coefficient reduction requires bigger nozzles than the ideal design.

It must be noted that the ideal and actual designs differ only in the outlet area size. All other dimensions remain the same. As in both designs the result is $A_f = A_g$, the two nozzles meet the optimum design condition of Formula 5.11, which means that the nozzle expansion is exact ($r_n = 1$).

However, thermodynamic properties undergo significant changes that finally produce a lower outlet velocity in nozzle 2. These changes are a consequence of the reheat produced by friction losses. Velocity of nozzle 2 is 10% lower than nozzle 1, Consequently, the former has 19% less of kinetic energy.

5.7.3. Nozzle performance curves

Having assumed that the flow is one-dimensional, it is feasible to visualize and assess a nozzle performance efficiency by means of curves of flow properties, versus abscissa x.

The performance curves are taken from a table of thermodynamic properties. This table is prepared by selecting an independent variable that finally allows to refer any thermodynamic property to its location along the x axis. As all thermodynamic and geometric properties are directly or indirectly, functions of the pressure and no thermodynamic property is a function of x, pressure is adopted as independent variable to create the performance curves table. After all thermodynamic properties and section areas have been calculated for a conical nozzle, it's possible to calculate their location, that is their abscissas x, with trigonometric formulas. If the nozzle is not conical, a specific formula of area sections must be derived to calculate the location x.

Table 5.8. Expansion ratios

Expansion ratios	Nomenclature and formula	Nozzle 1 Ideal design. $\varphi = 1$	Nozzle 2 Actual design. $\varphi < 1$
Specific volume	$X_v = \dfrac{V_2}{V_1}$	1.785	3.057
Velocity	$X_v = \dfrac{V_2}{V_t}$	1.434	1.290
Throat flow area. Actual value	$A_{tf} = A_t = G \cdot \dfrac{V_t}{V_t} \cdot 10^4$	6.93	cm²
Outlet flow area. Actual value	$A_{2f} = G \cdot \dfrac{V_2}{V_2}$	10.34	cm²
Flow	$X_f = \dfrac{A_{2f}}{A_{tf}}$	1.245	1.492
Geometric	$X_g = \dfrac{A_{2g}}{A_{tg}}$	1.245	1.492
Expansions ratio or outlet areas ratio	$r_A = \dfrac{X_g}{X_f} = \dfrac{A_{2g}}{A_{2f}}$	1.000	1.000

In conical nozzles there are two different trigonometric formulas of the abscissa x, one is valid for the convergent part and the other is applicable to the divergent part. Formula to be used is decided by comparing the current pressure with the nozzle critical pressure.

○ Formula for the convergent part. In this part pressure values are higher than critical pressure; $p_x > p_t$.

Formula 5.19. Abscissa of thermodynamic
properties in the convergent part

$$x = L_1 - \frac{D_x - D_t}{2 \cdot tg(\alpha_1)}$$

o   Formula for the divergent part. In the divergent part all
    pressures are lower than the critical pressure; $p_x < p_t$.

Formula 5.20. Abscissa of thermodynamic
properties in the divergent part

$$x = L_1 + \frac{D_x - D_2}{2 \cdot tg(\alpha_2)}$$

This example assumes that no normal shock wave happens
because nozzle discharge pressure is equal to design pressure.
However, in practice it's advisable to test the nozzle design and
assess potential impacts for discharge pressures other than
the design pressure. That's why a set of discharge pressures,
different to the design pressure, must be obtained from the
expected turbine operation.

An important conclusion to be drawn from this example is
the Thermodynamics – Geometry synergy. These sciences are
combined on the roots of the principle of mass continuity, and
that allows the optimum modelling of the nozzle shape and
dimensions. Nonetheless an optimum geometric design is not
all. As was said before, other requirements are also important
such as a good choice materials of high temperature resistance,
a strong structural design to endure internal pressures, and
a correct specification of inner walls treatment to minimize
friction losses.

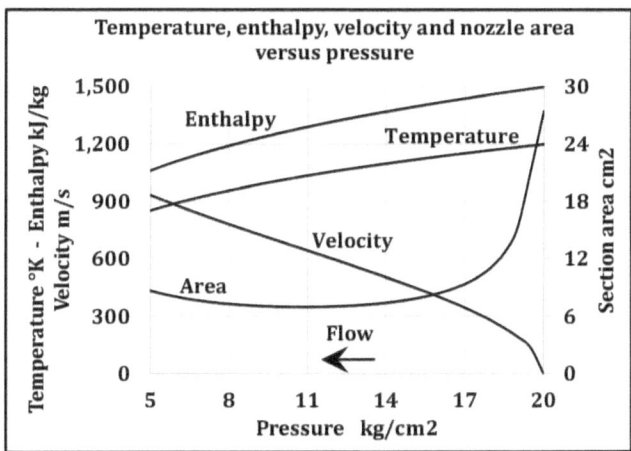

Figure 5.12. Thermodynamic properties versus pressure

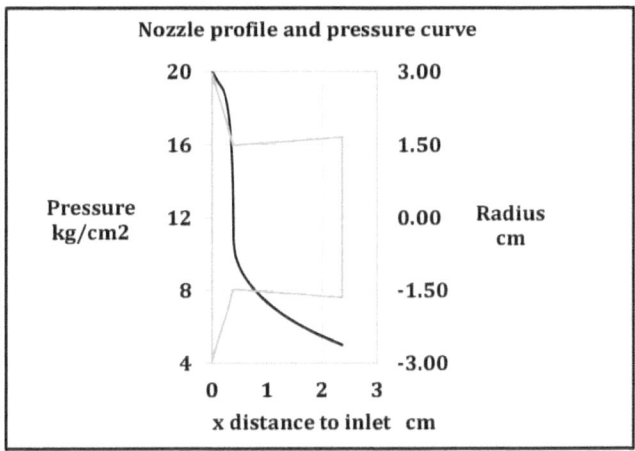

Figure 5.13. Pressure curve and nozzle profile

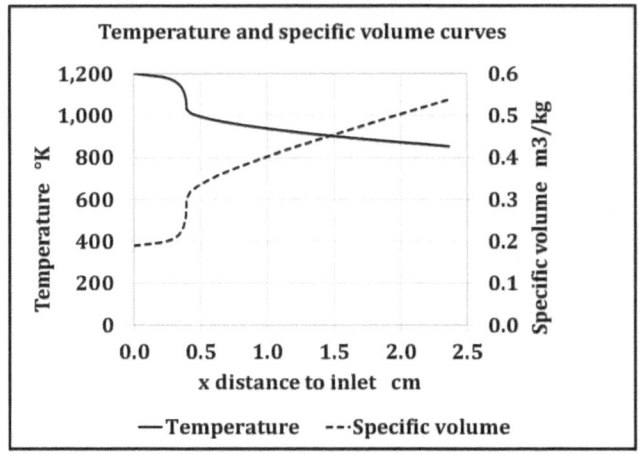

Figure 5.14. Temperature and specific volume curves

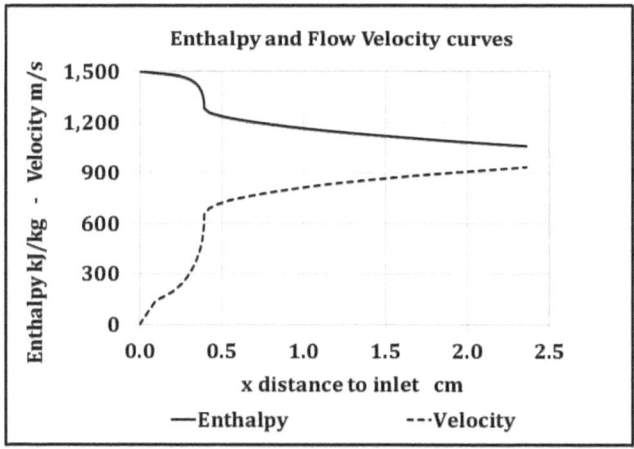

Figure 5.15. Enthalpy and velocity curves

5.8. Summary of most important nozzle aspects

With the example in section 5.7 this book has completed the basic theory and practice of gas and steam nozzles. Nonetheless it's important to close Chapter 5 with the following short summary of all theoretical and practical issues discussed in Part I.

o One-dimensional flow is generally adopted for the study of nozzles. It's not an exact assumption but it is a good approximation.

o Critical conditions at the throat produce a sonic flow at that section of CD nozzles. No sonic regime exists in C nozzles.

o A flow is choked when the nozzle pressure ratio p_t/p_1 is equal to or smaller than the pressure critical ratio.

o A choked flow velocity does not depend on the nozzle discharge pressure. It lingers constant and equal to the speed of sound. The same happens with all other thermodynamic properties.

o C type nozzles operate with pressure ratios higher than the critical pressure ratio. CD type nozzles operate with pressure ratios equal or smaller than the critical ratio.

o Velocities and mass flows of a not choked flow behave as the right descending branches of Figures 3.2 and 3.3 respectively.

o Mass flow is maximal when the pressure ratio equals, or is smaller than the critical ratio.

o It's not possible to increase the mass flow by reducing the discharge pressure p_2 of a choked flow.

o There are two different causes of heat released by friction; roughness and turbulence. Both effects are considered in only one number: friction factor.

o Discharge pressure value other than design specification impacts on friction factor by creating turbulent regimes.

o Overexpansion and underexpansion conditions produce external oblique shock waves, which reduce the kinetic energy of the flow. Impact of overexpansion is higher than underexpansion.

o When a CD nozzle operates at discharge pressures higher than the critical shock wave pressure (p_B), normal shock waves are formed inside the diverging portion. This does not happen in the converging nozzle.

o The impact of a normal shock wave is the kinetic energy reduction and a sudden pressure increase where the shock wave is located. This phenomenon is arguably a more harmful operating condition than the friction and turbulence consequences on the flow velocity and nozzle efficiency.

○ Nozzle dimension increase as velocity coefficient decreases.
○ The optimal nozzle shape and dimensions are identical to the flow geometric properties.

PART II
NOZZLE PROJECTS

Chapter 6

Gas Nozzle Project

This project requires to design a nozzle to be located at the discharge end of the combustion chambers of a gas turbine, whose power is 4,000 kW. The initial data for this project are taken from an actual turbine which drives an electric generator.

The design procedure to be explained may significantly differ from others because design procedures strongly depend on the initial data of each case. Anyway, the reading of this project allows to understand and organize any nozzle project, according to other dictates of a turbine's needs.

This project presentation does not show detailed calculations, but a sequence of formulas ready to carry out the appropriate calculations and the results. This is done in tables, each corresponding to an xlsx file, where formulas and results are presented in the appropriate sequence of calculation, just as it was done in the example of Chapter 5.

All tables and figures of this chapter are in file 6.1. Chapter 6. Gas nozzle project.xlsx. Refer to this file to review all calculations and figures shown in the text.

6.1. Project plan

The gas nozzle project includes the following sections:

○ Input data and physical constant
○ Critical properties and nozzle type identification
○ Thermodynamics properties

- o Energy properties
- o Flow properties
- o Geometric design
- o Expansion ratios
- o Performance table and curves

6.2. Input data and physical constants

All values shown in Input data and physical constants table, have been defined throughout this book. They are generally known and are easily obtained in other books, or in specialized sites on the internet.

It is important to remind that there are five constants that require a consistency verification. They are the following.

- o Universal gas constant R_u
- o Specific gas constant R
- o Gas molecular weight MW
- o Specific heat at constant pressure c_p
- o Specific heats ratio $k = c_p/c_v$

The expression used for their consistency verification is Formula 1.4.

The specified specific heat of Table 6.1 is the expected average value of combustion gases between inlet and outlet temperatures (1,300°K and 1,118°K respectively).

The atmospheric pressure at the turbine location is 1 kg/cm². The GT compressor has a 9.5:1 compression ratio. As pressure loss in the combustion chamber is considered negligible, then the air inlet pressure to the nozzle is equal to 9.5 kg/cm2.

The nozzle type is not imposed at this moment, but the upcoming calculations will prove that the pressure ratio is above critical and Consequently, the nozzle must be C type.

The velocity coefficient is 0.98, which is a typical low value for C type nozzles. The reason for this high velocity coefficient is that convergent nozzles are short and therefore, friction, turbulence and pressure losses are low. In fact, it is usual to neglect friction loss in C type nozzles or in the convergent part of CD nozzles. However, in this example heat loss is not disregarded.

If actual discharge pressure is not equal to design pressure, it will surely be necessary to redo the calculations until final velocity coefficient matches with its initially assumed value. This is a straightforward task in any specialized software or in a spreadsheet. A good example of this procedure is in the steam nozzle project of Chapter 7.

On the basis of previous considerations made in respect of the negligible inlet velocity versus the outlet velocity, it's assumed that $V_1 = 0$. However, as in GT nozzles the combustion chamber gets its air from a compressor, it is recommended to verify this point, because the outlet velocity of turbine compressors can reach 150 m/s, and that velocity might impact the combustion environment. It is for this reason that combustion chambers generally have an inlet diffuser shape in order to reduce air speed. The outgoing end of the chamber is a nozzle, to increase the jet kinetic energy. Thus the combustion chamber has an almost ovoid shape (or goose egg shape), whose pointed ends are the diffuser inlet and the nozzle discharge.

Table 6.1. Input data and physical constants

Thermodynamic input data			
	Designation	**Value**	**Units**
Fluid	Combustion gases		
Pressure input	p1	11.00	kg/cm2 a
Discharge pressure	p2	6.00	kg/cm2 a
Input gases temperature	T1	1,300	°K
Mass flow	Gm	2.00	kg/s
Input velocity	V1	0	m/s
Velocity coefficient	φ	0.980	
Geometric input data			
Convergent angle α1	45	°	
Divergent angle α2	5	°	
Inlet area/Outlet area	9		
Physical constants			
Mechanical equivalent of heat	J	102.00	kg.m/kJ
Gravity acceleration	g	9.81	m/s2
Universal constant of ideal gas	Ru	8.314	kJ/|kmol.°K|
Specific heats ratio. k = cp/cv	k	1.330	Ver Tabla 1
Specific constant R of combustion gases	R	0.3310	kJ/|kg . °K|
Average specific heat. T range 1,160°K to 1,300°K	cp	1.3339	kJ/|kg . °K|
Velocity coefficient	aV	3.347	See Table 3.1
Mass flow coefficient	aG	2.107	See Table 3.1
Air viscosity	μ0	1.716E-05	mass kg/|m.s|
Reference temperature of μ0	Tref	323	°K
Sutherland temperature for air	Tu	110	°K

6.3. Critical properties and nozzle type identification

Critical property that identify the applicable nozzle type is the pressure at the throat. As was discussed earlier, if nozzle pressure ratio r_{pn} is higher than critical pressure ratio r_{pc}, nozzle must be C type, otherwise it has to be CD type. See Table 6.2.

Table 6.2. Critical properties and nozzle identification

Properties	Formulas and results
Nozzle pressure ratio	$r_{pn} = \dfrac{p_2}{p_1} = 0.545$
Critical pressure ratio	$r_{pc} = \left(\dfrac{2}{k+1}\right)^{\frac{k}{k-1}} = 0.540$
Critical temperature ratio	$r_{tc} = \dfrac{T_c}{T_1} = \dfrac{2}{k+1} = 0.858$
Critical specific volumen ratio	$r_{vc} = \dfrac{V_c}{V_1} = \left(\dfrac{k+1}{2}\right)^{\frac{1}{k-1}} = 1.589$
Critical pressure kg/cm² a	$p_c = r_{pc} \cdot p_1 = 5.94$
Critical Temperature, °K	$T_c = r_{pc} \cdot p_1 = 1,116$
Critical specific volume, m³/kg	$V_c = r_{vc} \cdot V_1 = 0.634$
Nozzle type	as $p_2 > p_c$ nozzle must be C type

In a converging nozzle, critical values of velocity and pressure are not located within the nozzle because the sonic regime can't be achieved inside it. In some technical publications it is said that the critical pressure is outside the nozzle. This expression is not exact and could create some confusion because the outlet pressure is given by the environment pressure. In this case the outgoing flow penetrates in the environment and moves within it until its pressure equals the environment pressure.

6.4. Thermodynamics properties

In Table 6.3 thermodynamic properties at the nozzle inlet are calculated in the third column and properties at the outlet are in the fourth column. As properties values at the outlet are dependent on their counterparts at the inlet, calculation of any property always starts at the nozzle inlet.

Table 6.3. Thermodynamic properties

Description	Inlet Formulas and results	Outlet Formulas and results
Pressure kg/cm2 a	$p_1 = \text{Initial data} = 11.0$	$p_2 = \text{Initial data} = 6.0$
Temperature °K	$T_1 = \text{Initial data} = 1{,}300$	$T_{2ideal} = T_1 \cdot \left(\dfrac{p_2}{p_1}\right)^{\frac{k-1}{k}} = 1{,}118$
Specific volume m³/kg	$v_1 = \dfrac{J \cdot R \cdot T_1}{p_1 \cdot 10^4} = 0.462$	$v_2 = v_1 \cdot \left(\dfrac{p_1}{p_2}\right)^{\frac{1}{k}} = 0.629$
Specific gravity kg/m³	$sg_1 = \dfrac{1}{v_1} = 2.165$	$sg_2 = \dfrac{1}{v_2} = 1.589$
Actual temperature °K	$T_{2actual} = T_2 + \dfrac{q_f}{c_p} = 1{,}126$	This temperature includes the ΔT produced by the reheat q_f, which incorporates to the discharge gas

Table 6.3 is the cornerstone of nozzle design. All other calculations are directly or indirectly related to results returned by this Table. From this fact comes the importance of the input data

quality, because they are essentials to a proper thermodynamic calculation. Calculations in Table 6.3 only use Formulas 1.3 and 1.9.

Inlet temperature T_1 is generally known from manufacturer's information, but discharge temperature T_2 is calculated with the isentropic relations of Formula 1.9. Inlet specific volume is calculated with equation of state formula and discharge specific volume is also obtained with the isentropic relations. The gas expansion is evidenced by a specific volume at the inlet, which is smaller than at the outlet

6.5. Energy properties

As nozzles are a turbine component, the former determines nozzle's technical specifications. Typically required values are: temperature, flow velocity or kinetic energy of fluid delivered to the first rotor and mass flow.

Enthalpies, available energy, kinetic energy and nozzle friction losses are calculated in Table 6.4. Arguably, the nozzle is one of the most efficient components of a gas turbine because flow is accelerated from almost zero velocity at the inlet, to near sonic (C type nozzles) or supersonic (CD type nozzles) regimes at the outlet, with little energy's loss. In this example, energy loss is 4% of total available energy in the combustion gas.

Calculation starts with enthalpy values at the nozzle inlet and outlet sections. Applicable formulas and results are shown in the row "Enthalpy. Ideal", where calculation of discharge enthalpy is done under ideal conditions; that is with no friction between nozzle inner wall and gas flow. That's why h_2 is subscripted ideal. Difference between enthalpies at the inlet and at the outlet sections represents total available energy to accelerate the flow. In this example, this energy is only 6% of input enthalpy.

Total energy is converted into mechanical energy in the turbine, though not entirely because of losses in the conversion process. In turbine technology total available energy is also named total enthalpy.

Friction heat loss is calculated in row Reheat. This heat loss is released at the expense of total available energy and after it's incorporated to discharge ideal enthalpy, which results in a higher value.

The jet discharge velocity is calculated with the enthalpies difference between nozzle inlet and outlet sections, as shown in rows "Ideal velocity" and "Actual velocity". Flow velocity reduction from 696 m/s to 682 m/s is a direct consequence of friction.

The calculation sequence follows with kinetic energies. The most important energy is that of the discharge flow because the turbine will drive the blades wheel at the expense of this energy. The outflow ideal kinetic energy is calculated in row "Kinetic energy ideal" and observance of the principle of energy conservation is verified in Row "Total Energy". To do this, ideal kinetic energy and enthalpy are summed and shown in the appropriate column. This value of the total energy must be constant throughout the nozzle. The same calculation is done in rows "Total energy. Actual" and "Total Energy. Validation", under actual conditions. As heat loss is included in the total energy value, this energy equals the total ideal energy. The kinetic energy E_{k2} value minus different losses produced in the turbine, will represent the shaft mechanical energy of the turbine.

Finally, in row "Conversion energy", this parameter is calculated. As was said in Chapter 2 this indicator should be construed as the conversion efficiency of the total heat energy h_1 at the nozzle inlet into kinetic energy at the outlet. The enthalpy h_1 is also known as "stagnant enthalpy", which is supposed to be the gas actual state at zero velocity.

Converging nozzles have very low conversion efficiency unlike CD nozzles where the diverging portion adds a further increase in speed and of course, also in the outflow kinematic energy.

6.6. Flow properties

Flow properties define ideal power delivered by one nozzle to the turbine's blades wheel, where discharge jet power is instrumental to design or assess nozzle performance.

Further calculations are: outlet specific mass flow, sound velocity and Mach number. As this is a C nozzle, the Mach number at the outlet is lower than 1. In this case the Mach number is 0.97. Although this number is pretty close to unity, flow is not choked.

Table 6.4. Energy properties

Property	Inlet Formulas and results	Outlet Formulas and results
Enthalpy. Ideal kJ/kg	$h_{1ideal} = h_1$ $= c_{p1} \cdot T_1 = 1{,}734$	$h_{2ideal} = c_{p2} \cdot T_{2ideal}$ $= 1{,}492$
Total available energy (TAE) kJ/kg	$E_{a\,ideal}$ $= h_1 - h_{2ideal}$ $= 242$	NA
Nozzle efficiency %	$e_n = \varphi^2 \times 100$ $= 96.04\%$	NA
Reheat kJ/kg	$q_f = \left(1 - e_n\right)$ $\cdot \left(h_1 - h_{2ideal}\right) = 9.6$	$q_{f\%} = \dfrac{q_f}{E_{a\,ideal}} \times 100$ $= 3.96\%$ of TAE
Actual enthalpy kJ/kg	See first row $h_1 = 1{,}734$	$h_2 = h_{2ideal} + q_f = 1{,}502$
Ideal velocity m/s	$V_{1ideal} = 0$	V_{2ideal} $= \sqrt[2]{2{,}001 \cdot \left(h_1 - h_{2ideal}\right)}$ $= 696$
Actual velocity m/s	$V_1 = V_{1ideal} = 0$	$V_2 = \sqrt[2]{2{,}001 \cdot \left(h_1 - h_2\right)} = 682$
Kinetic energy. Ideal kJ/kg	$E_{k1ideal} = 0$	$E_{k2ideal} = \dfrac{V_{2ideal}^2}{2 \cdot g \cdot J} = 242$
Total energy kJ/kg	$E_{t1ideal}$ $= h_1 + E_{k1ideal}$ $= 1{,}734$	$E_{t2ideal} = h_{2ideal} + E_{k2ideal}$ $= 1{,}734$

Kinetic energy. Actual kJ/kg	$E_{k1} = 0$	$E_{k2} = \dfrac{V_2^2}{2 \cdot g \cdot J} = 232$
Conversion efficiency %	$e_c = \dfrac{E_{k2}}{h_1} \times 100 = 13.41\%$ of h_1	

Table 6.5. Flow properties

Property	Inlet Formulas and results	Outlet Formulas and results
Ideal mass flow kg/s	0.00	$G = \dfrac{A_2 \cdot V_{2ideal}}{V_2} = 2.04$
Actual mass flow kg/s	0.00	$G = \dfrac{A_2 \cdot V_2}{V_2} = 2.00$
Specific mass flow kg/|cm².s|	$g_1 = \dfrac{G}{A_1} = 0.00$	$g_2 = \dfrac{G}{A_2} = 0.1089$
Speed of sound m/s	$V_s = \sqrt[2]{g \cdot k \cdot J \cdot R \cdot T_t}$ $= 757$	$V_s = \sqrt[2]{g \cdot k \cdot J \cdot R \cdot T_2}$ $= 704$
Mach number	0.00	$M_2 = \dfrac{V_2}{V_s} = 0.97$
Discharge jet power kW or kJ/s	NA	$P = G \cdot E_{k2} = 465$

6.7. Geometric design

Conical C type nozzles have four geometric properties that fully define the nozzle shape and dimensions.

○ L_1: nozzle length

o D_1: inlet diameter
o D_2: outlet diameter
o α_1: convergence angle

These four variables are related by the following trigonometric formula.

Formula 6.1. Convergent part length

$$L_1 = \frac{D_1 - D_2}{2 \cdot tg\alpha_1} \qquad |cm|$$

The calculation procedure is as follows.

o A_2 is calculated according to the principle of continuity. D_2 is a direct calculation from the A_2 value.
o   Thermodynamics does not provide any formula for the other three dimensions of Formula 6.1, therefore, at least two of them must be prefixed. This is done according to the designer's experience, space availability, etc. The converging angle α_1 is usually prefixed in the range of 30° to 75. In this case it was specified $\alpha_1 = 45°$.
o   As the inlet to outlet ratio is prefixed, A_1 = Areas ratio × A_2. The corresponding diameter D_1 is also a direct calculation like in the outlet diameter D_2.
o   The nozzle length is then calculated with the trigonometric
o   Formula6.1.

Table 6.6. Geometric design

Property	Inlet. Formulas and results	Outlet. Formulas and results
Nozzle angles °	α_1: This angle is prefixed, typically in the range of 30° a 75°. For this case it was adopted = 45°.	α_2: Only for CD nozzles. This angle is typically prefixed in the range of 5° a 15°. Not applicable in this case.
Inlet area cm²	$A_1 = \text{Areas ratio} \times A_2 = 166.1$	$A_2 = \dfrac{G \cdot v_2}{V_2} = 18.5$
Inlet and outlet diameters cm	$D_1 = \sqrt[2]{\dfrac{4 \times A_1 \times 10^4}{\pi}}$ $= 14.54$	$D_2 = \sqrt[2]{\dfrac{4 \times A_2 \times 10^4}{\pi}}$ $= 4.85$
Nozzle total length, cm	$L_1 = \dfrac{D_1 - D_2}{2 \times tg(\alpha_1)} = 4.85$	
Nozzle total volume cm³	$Vol_t = \pi \cdot \dfrac{L_1}{12} \cdot \left(D_1^2 + D_2^2 + D_1 \cdot D_2\right) = 387.6$	

6.8. Expansion ratios

The expansion ratios are non-dimensional numbers that measure how much some properties have changed between the nozzle inlet or throat and the outlet. They are calculated by dividing an outlet property value over the same property value at the throat. As in C type nozzles the outlet and throat sections are the same, all expansion ratios are equal to 1. The usual expansion ratios applicable in practice are the following.

o   Specific volume expansion. This ratio measures the gas expansion between the throat and the outlet sections. See first row of Table 6.7

o   Velocity expansion. This ratio compares the outflow velocity with the throat flow velocity. It's an indirect measure of the gas acceleration between the throat and the outlet. See second row of Table 6.7.

o   Flow areas expansion. This ratio indicates the increase of the flow cross section between the throat and the outlet. See third row of Table 6.7.

o   Geometric area expansion. This is a pure geometric (or constructive) indicator of the difference between the throat and the outlet nozzle sections. See fourth row of Table 6.7.

Table 6.7. Expansion ratios

Property	Formulas and results	Notes
Specific volume expansion ratio	$X_v = \dfrac{V_2}{V_1} = 1.000$	
Velocity expansion ratio	$X_{Velocity} = \dfrac{V_2}{V_g} = 1.000$	As the outlet and throat in a type C nozzle coincide, all expansion ratios are forcefully equal to 1.
Flow expansion ratio	$X_f = \dfrac{A_{2f}}{A_{tf}} = \dfrac{V_g}{v_g} \cdot \dfrac{v_2}{V_2} = 1.000$	
Geometric expansion ratio	$X_g = \dfrac{A_2}{A_t} = 1.000$	

o   The aftermath of Table 6.7 is that expansion ratios don't provide any information about the convergent nozzle performance. This is not the case in CD type nozzles, where these ratios are helpful to calculate velocity coefficients, as was seen in section 5.4.

6.9. Performance table and curves

The performance table and curves generation must be done according to the same formulas and procedure used in section 5.7. Therefore, all considerations and clarification made in that section are also valid for this case.

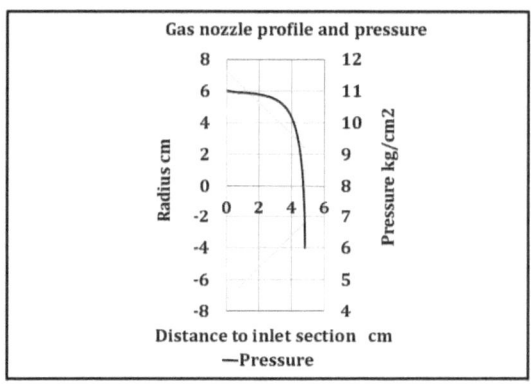

Figure 6.1. Nozzle profile and pressure curve

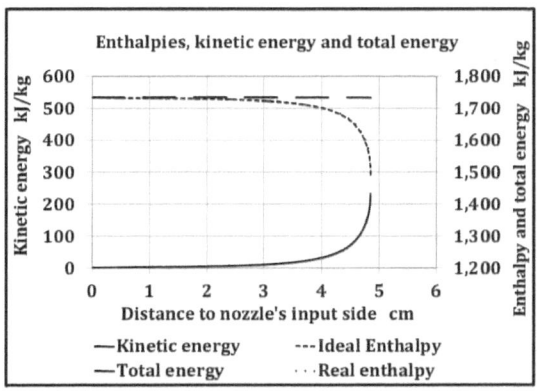

Figure 6.2. Enthalpies, kinetic energy and total energy versus distance to inlet

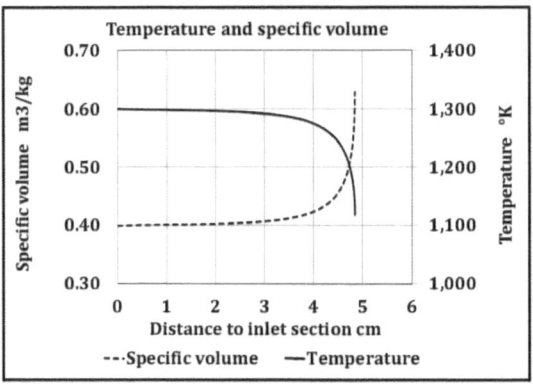

Figure 6.3. Temperature and specific volume curves

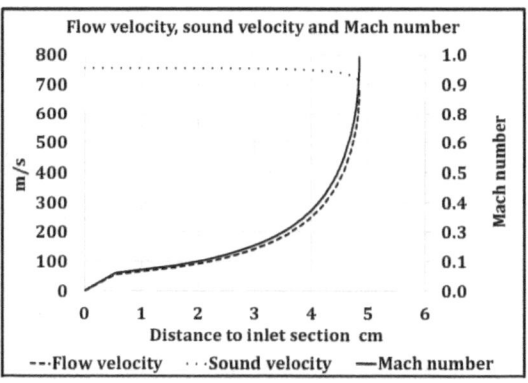

Figure 6.4. Flow velocity, sound velocity and Mach number curves

An important conclusion of Figures 6.1 to 6.4 is that all state variables exhibit a sudden change next to the nozzle outlet. It means that almost no thermodynamic variations occur in most of the first half of the nozzle. However, as flow approaches to the outlet constriction a fast pressure loss occurs, anticipating that any flow throttling produces sudden changes to the fluid's thermodynamic state. This phenomenon is especially important in CD nozzles because it produces velocities higher than speed of sound.

CHAPTER 7

STEAM NOZZLE PROJECT

As in many other engineering applications there is no single method for designing a nozzle and therefore, an example with universal validity can't be provided. Nonetheless, a nozzle design procedure and its assessment under conditions different to design specifications are presented. This example helps to understand how to use in practice the concepts and formulas about friction impact and shock waves discussed in this book.

This project refers to a very common situation, where engineers must analyze, generate alternatives, compare and then implement a solution or... do nothing, provided that everything is working properly. In this case, for operative reasons not connected with the nozzles, a steam turbine specialist is requested to change the nozzles discharge pressure from 2 to 10 kg/cm². The questions are:

○ Is this change technically feasible?
○ Will the turbine performance be impacted if this change is implemented?
○ If so, do solutions exist to prevent undesirable consequences?

Comments to this project:

○ The design is done according to the optimum design criterion discussed in section 5.2, which is: $A_{f2} = A_{g2}$.
○ The design procedure explained in this chapter is completed by a sensitivity analysis of nozzle efficiency versus discharge pressure in section 7.9.

○ Since the case to study is a superheated steam nozzle, transformations calculation is made using ideal gas formulas. Of course, enthalpies are calculated with the Mollier formula.
○ Files including detailed calculation are
  ○ 7.1. Chapter 7. Case 2. Steam Nozzle assessment.xlsx
  ○ 7.2. Chapter 7. Case 1. Steam Nozzle design.xlsx
  ○ 7.3. Chapter 7. Case 3. Steam Nozzle assessment.xlsx
  ○ 7.4. Chapter 7. Case 4. Steam Nozzle assessment.xlsx
  ○ 7.5. Chapter 7. Case 5. Steam Nozzle assessment.xlsx
  ○ 7.6. Chapter 7. Case 6. Steam Nozzle assessment.xlsx
  ○ 7.8. Chapter 7. Shock wave calculation.xlsx

○ File 7.2 is the nozzle design calculation. Its output is formed of thermal properties at the outlet and the nozzle geometric design as well. Therefore, the applicable design chart to File 7.2 is the flow chart of Figure 5.10.
○ Files 7.1, 7.3, 7.4, 7.5 and 7.6 are nozzle assessments for different discharge pressures. None of them include a geometric design procedure because it's already done. Consequently, the nozzle geometry is part of the Data Input. See flow chart of Figure 7.1.
○ The second digit of file numbers is coincident with the discharge pressure value of that case, in kg/cm². Exception was made with file 7.1 whose outlet pressure is 1.2 kg/cm².
○ For pressures other than design pressure, the initially adopted velocity coefficient is very unlikely to equal the value returned by block named Formula $\varphi(r_A)$. This value is the calculated velocity coefficient φ_c. The initial velocity coefficient is designed as φ_i. But as the final result requires that both are the same, their difference must be set to zero, that is:

$$\Delta\varphi = \varphi_i - \varphi_c = 0$$

This is done manually with the Goal Seek instruction of the spreadsheet. This instruction recalculates the complete model until $\Delta\varphi$ equals zero, by changing the initial velocity coefficient φ_i. The result is the definitive velocity coefficient. See this feedback loop in grey colored blocks of Figure 7.1. In summary: after all data is introduced in the Input Data block, the Properties Mathematical Model does not show the final output. This is only obtained after the Goal Seek instruction of the spreadsheet is manually launched as explained.

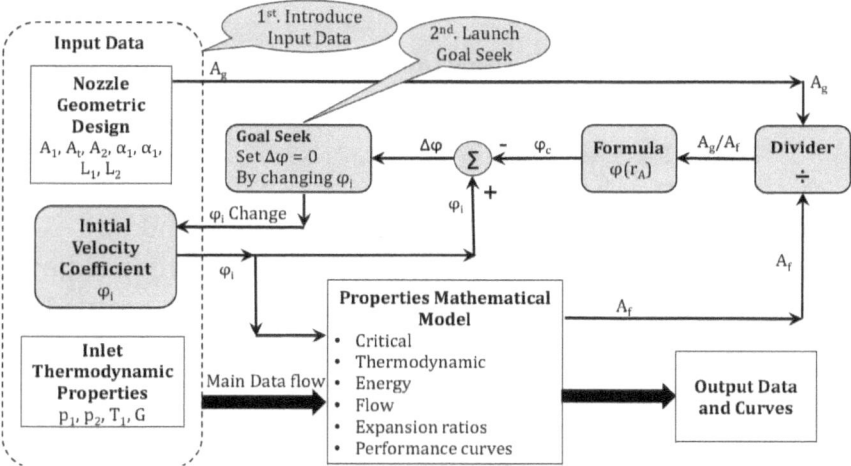

Figure 7.1. Nozzle assessment flow chart

7.1. Project technical specifications

The project presented in this chapter aims to determine the following specifications for designing purposes. Calculation of the following properties is specified:

○ Critical properties and the nozzle type.
○ Thermodynamic properties at the inlet, throat and outlet.
○ Flow energetic properties.
○ Kinetic energy of discharge flow
○ Ideal and actual enthalpy of discharge flow
○ Friction losses
○ Nozzle dimensions for maximum efficiency.

- o   Design a conical nozzle:
    - o   Inlet diameter
    - o   Throat diameter
    - o   Discharge diameter
    - o   Divergent portion's length
    - o   Nozzle's total length
- o   Expansion ratios (X_v, X_{Vel}, X_f and X_g)
- o   Performance curves for design discharge pressure versus axial location x.
- o   Nozzle efficiency for operating conditions other than design specifications.
    - o   Case 1. p2 = 1.2 kg/cm2 a
    - o   Case 2. p_2 = 5.0 kg/cm² a
- o   Velocity coefficient and efficiency for cases a) and b).
- o   Pressure curve resulting from a shock wave produced by a pressure equal to 10 kg/cm² and located at a distance x = 1.75 cm from the inlet.

Table 7.1. Design condition. Case 2. Input data and physical constants

Thermodynamic input data			
Property	**Designation**	**Value**	**Units**
Fluid	Superheated steam		
Atmospheric pressure	patm	1.00	kg/cm2 a
Compressor input pressure	p1	15.00	kg/cm2 a
Discharge design pressure	p2	2.00	kg/cm2 a
Input gases temperature	T1	873	°K
Mass flow	Gm	0.50	kg/s
Input velocity	V1	0	m/s
Initial velocity coefficient	φi	0.971	
Geometric input data			
Convergent angle α1	50 °		0.873
Divergent angle α2	5 °		0.087
Area 1/Area t ratio	4.00	D1/Dt	2.00
Physical constants			
Mechanical equivalent of heat	J	102.00	kg.m/kJ
Gravity acceleration	g	9.81	m/s2
Universal constant of ideal gas	Ru	8.314	kJ/|kmol.°K|
Molecular weight of steam	PM	18.46	kg/kmol
Specific constant R of steam	R	0.4504	kJ/|kg . °K|
Specific heats ratio. k = cp/cv	k	1.300	Ver Tabla 1
Specific heat at constant pressure	cp	1.952	kJ/|kg . °K|
Velocity coefficient	aV	3.330	See Table 3.1
Mass flow coefficient	aG	2.090	See Table 3.1

For sections 7.2 to 7.8 see file 7.1. Chapter 7. Case 2. Steam Nozzle design.xlsx.

7.2. Case 2. Critical properties and nozzle type identification

Table 7.2. Case 2. Critical properties and nozzle type

Property	Formula and result
Nozzle pressure ratio	$r_{pn} = \dfrac{p_2}{p_1} = 0.133$
Critical pressure ratio	$r_{pc} = \left(\dfrac{2}{k+1}\right)^{\frac{k}{k-1}} = 0.546$
Critical temperature ratio	$r_{tc} = \dfrac{T_c}{T_1} = \dfrac{2}{k+1} = 0.870$
Critical specific volumen ratio	$r_{vc} = \dfrac{v_c}{v_1} = \left(\dfrac{k+1}{2}\right)^{\frac{1}{k-1}} = 1.593$
Critical pressure kg/cm2 a	$p_c = r_{pc} \cdot p_1 = 8.19$
Critical temperature °K	$T_c = r_{pc} \cdot p_1 = 759$
Critical specific volume m³/kg	$v_c = r_{vc} \cdot v_1 = 0.426$
Nozzle type	As $r_{pn} < r_{pc}$, nozzle type is CD

7.3. Case 2. Thermodynamic properties

Table 7.3. Case 2. Thermodynamic properties

Property	Inlet	Throat	Outlet
Pressure Kg/cm² a	$p_1 = 15.00$	$p_t = 8.19$	$p_2 = 2.00$

Condensation temperature: $T_{cx} = 100.86 \times p_x^{0.2477} + 273$			
	$T_{cond1} = 470$	$T_{condt} = 443$	$T_{cond2} = 393$
Temperature °K	$T_1 = 873$ Table 7.1	$T_t = 759$ Table 7.2	T_2 $= T_1$ $\cdot \left(\dfrac{p_2}{p_1} \right)^{\frac{k-1}{k}}$ $T_2 = 548$
Specific volume m³/kg	$v_1 = \dfrac{J \cdot R \cdot T_1}{p_1 \cdot 10^4}$ $v_1 = 0.267$	$v_t = v_1 \cdot \left(\dfrac{p_1}{p_c} \right)^{\frac{1}{k}}$ $v_t = 0.426$	$v_2 = v_1 \cdot \left(\dfrac{p_1}{p_2} \right)^{\frac{1}{k}}$ $v_2 = 1.260$

Notes:

○ As in this case p_2 is the specified design pressure, in this book it's also represented as p_D.

○ As condensation temperatures are smaller than steam temperatures, ideal gas formulas are applicable in this case forwater drops are not likely to be formed in the steam mass.

7.4. Case 2. Energy properties

Table 7.4. Case 2. Energy properties

Property	Inlet	Throat	Outlet

Enthalpies and efficiency

Steam enthalpy. Superheated steam			
$$h = 3.186 \times \left[0.47 \times (T-273) - \frac{201.96}{\left(\frac{T}{100}\right)^{\frac{10}{3}}} \times p - \frac{0.2248 \times 10^{12}}{\left(\frac{T}{100}\right)^{14}} \times \left(\frac{p}{100}\right)^3 + 595 \right]$$			

Property	Inlet	Throat	Outlet
Ideal enthalpy kJ/kg	$h_{1\,ideal} = 3,662$	$h_{t\,ideal} = 3,439$	$h_{2\,ideal} = 3,027$
Nozzle efficiency %	φ recalculated $= 0.970$	$e_n = \varphi^2 \times 100 = 94.09\%$	
Reheat kJ/kg	$q_{f1} = 0$	q_{ft} $= (1 - e_n)$ $\cdot (h_1 - h_{t\,ideal})$ $= 13.2$	q_f $= (1 - e_n)$ $\cdot (h_1 - h_{2ideal})$ $= 37.5$
Actual enthalpy kJ/kg	$h_1 = h_{1ideal}$ $= 3,662$	$h_t = h_{t\,ideal}$ $= 3,439$	$h_2 = h_{2ideal} + q_f$ $= 3,064$
Available energy kJ/kg	$E_{a\,ideal} = h_1 - h_{2\,ideal} = 635$	$e_c = \dfrac{E_{k2}}{h_1} \times 100 = 16.3\%$	

Velocities

Ideal velocity m/s	$V_{1\,ideal} = 0$	$V_{t\,ideal} = V_{st}$ $= \sqrt[2]{g \cdot k \cdot J \cdot R \cdot T_t}$ $= 668$	V_{2ideal} $= \sqrt[2]{2 \cdot g \cdot J \langle (h_1 - h_{2id})}$ $V_{2ideal} = 1{,}127$
Actual velocity m/s	$V_1 = 0$	$V_t = V_{t\,ideal}$ $= 668$	$V_2 = \sqrt[2]{2 \cdot g \cdot J \cdot (h_1 - h_2)}$ $V_2 = 1{,}094$

Kinetic and total energies

Ideal kinetic energy kJ/kg	$E_{k1ideal} = 0$	$E_{kt\,ideal} = \dfrac{V_{t\,ideal}^2}{2 \cdot g \cdot J}$ $= 230$	$E_{k2ideal}$ $= \dfrac{V_{2ideal}^2}{2 \cdot g \cdot J}$ $= 635$
Actual E_k kJ/kg	$E_{k1} = 0$	$E_{kt} = \dfrac{V_t^2}{2 \cdot g \cdot J} = 223$	$E_{k2} = \dfrac{V_2^2}{2 \cdot g \cdot J} = 598$
Total actual energy kJ/kg	E_{1total} $= h_1 + E_{k1}$ $= 3{,}662$	$E_{t\,total} = h_1 + E_{k1}$ $= 3{,}662$	E_{2total} $= h_2 + E_{k2}$ $= 3{,}662$

Similarly, to the example of Chapter 6, outlet temperature was also re-calculated for information purposes only, because friction impact is already included in the actual h_2 of Table 7.4. Actual enthalpy is $h_2 = h_{2\,ideal} + q_f = 3{,}099$ kJ/kg. This value was introduced in enthalpy formula of superheated steam. Then actual temperature T_2 was cleared by means of goal seek instruction in tab named Table 7.3. The result is $T_{2actual} = 567°K$, which is 3.5% higher than the ideal value (548°K).

7.5. Case 2. Geometric design of conical nozzle

Table 7.5. Case 2. Geometric design

Units: D|cm| A|cm²| Vol|cm³|

Inlet	Throat	Outlet
Adopted: $\alpha_1 = 45°$	NA	Adopted: $\alpha_2 = 5°$
A_1 = areas ratio $\times A_t$ = 13.15	$A_t = \dfrac{G \cdot v_t}{V_t} = 3.29$	$A_2 = \dfrac{G \cdot v_2}{V_2} = 5.76$
$D = \sqrt{\dfrac{4 \cdot A}{\pi}} = 4.09$	$D_t = \sqrt[2]{\dfrac{4 \cdot A_g}{\pi}} = 2.05$	$D_2 = \sqrt[2]{\dfrac{4 \cdot A_2 \cdot 10^4}{\pi}} = 2.71$
$L_1 = \dfrac{D_1 - D_t}{2 \cdot tg(\alpha_1)} = 0.85$		$L_2 = \dfrac{D_2 - D_t}{2 \cdot tg(\alpha_2)} = 3.96$
$L_{total} = L_1 + L_{throat} + L_2$	4.81	
$Vol_1 = \dfrac{\pi \cdot L_1}{12} \cdot \left(D_1^2 + D_t^2 + D_1 \cdot D_t\right)$ = 6.3		$Vol_2 = \dfrac{\pi \cdot L_2}{12} \cdot \left(D_2^2 + D_t^2 + D_2 \cdot D_t\right)$ = 17.5
$Vol_{total} = Vol_1 + Vol_2 = 23.8$		

7.6. Case 2. Flow properties

Table 7.6. Case 2. Flow properties

Property	Inlet	Throat	Outlet
Actual mass flow kg/s	$G_1 = \dfrac{A_1 \cdot V_1}{v_1 \cdot 10^4} = 0.00$	$G_t = \dfrac{A_t \cdot V_t}{t \cdot 10^4} = 0.50$	$G_2 = \dfrac{A_2 \cdot V_2}{v_2 \cdot 10^4} = 0.50$
Specific mass flow m³/kg	$g_{s1} = \dfrac{G_1}{A_1} = 0.000$	$g_{st} = \dfrac{G_m}{A_t} = 0.1520$	$g_{s2} = \dfrac{G_m}{A_2} = 0.0868$
Sound velocity m/s	$V_{s1} = k \cdot J \cdot g \cdot R \cdot T_1 = 715$	$V_{st} = k \cdot J \cdot g \cdot R \cdot T_t = 667$	$V_{s2} = k \cdot J \cdot g \cdot R \cdot T_2 = 567$
Actual Mach number	$M_1 = \dfrac{V_1}{V_{s1}} = 0.000$	$M_t = \dfrac{V_t}{V_{st}} = 1.001$	
Discharge jet power kW	$P_2 = G_m \cdot E_{k2} = 299$		

7.7. Case 2. Expansion ratios and efficiency validation

Table 7.7 demonstrates that the flow and geometric expansion ratios are equal. This equality means that the flow shape is fitted to the nozzle shape, hence, turbulence is minimal.

Table 7.7. Case 2. Expansion ratios and efficiency validation

Property	Formula and results	
Specific volume and velocity expansion ratios	$X_v = \dfrac{v_2}{v_1} = 2.957$	$X_{Velocity} = \dfrac{V_2}{V_t} = 1.688$
Flow and geometric area expansion ratios	$X_f = \dfrac{A_{2f}}{A_{tf}} = \dfrac{V_t}{v_t} \cdot \dfrac{v_2}{V_2} = 1.751$	$X_g = \dfrac{A_2}{A_t} = 1.751$
Outlet areas ratio A_g/A_f	$r_A = \dfrac{X_g}{X_f} = \dfrac{A_{2f}}{A_{tf}} = 1.000$	Formula 5.17 results times 0.970 validates the initially adopted φ. $\varphi_{initial} = 0.970$ $\varphi_{recalculated} = 0.970$ $e_n = 94.09\%$

Conclusion: as the geometric and flow expansion ratios are the same, the expansion type is exact. Consequently, the velocity coefficient is the maximum resulting from the Steinmetz curve, downwardly displaced 0.03 points. This displacement is just an assumption adopted for this example. See chart of Figure 5.9.

7.8. Case 2. Performance curves

Based on the results of the above calculations, performance curves are plotted versus abscissa x, in Figures 7.2 to 7.5.

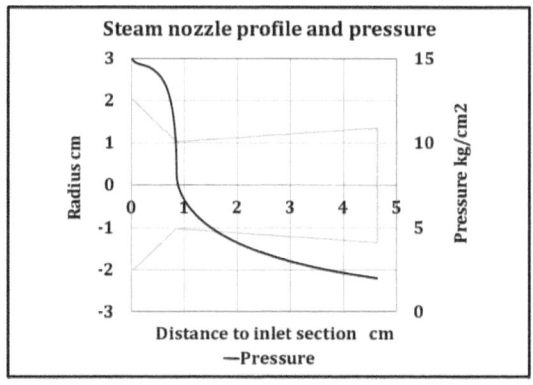

Figure 7.2. Nozzle profile and pressure curve

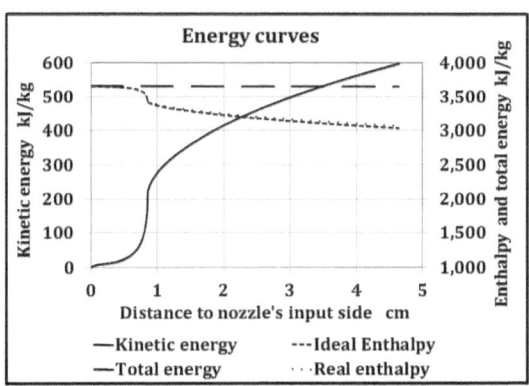

Figure 7.3. Energy curves

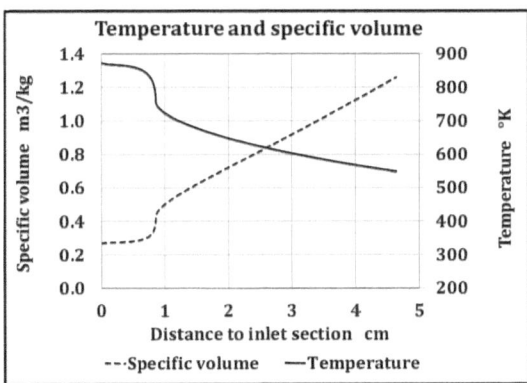

Figure 7.4. Temperature and specific volume curves

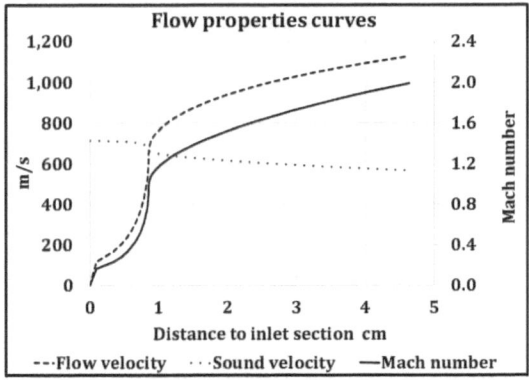

Figure 7.5. Velocities and Mach number curves

7.9. Efficiency assessment under operating conditions other than design specifications

This section evaluates nozzle efficiency for two different values of discharge pressure; one is lower than design pressure and the other is higher. Only the sheet of expansion ratios and flow properties will be produced below because the methodology used for all previous tableshas already been explained.

In the book's site five files have been uploaded. All of them are performance assessments of the nozzle designed in this chapter, for different discharge pressure. Only cases 1 and 5 are presented in this section. Anyway cases 3, 4 and 6 have been uploaded and can be consulted. Refer to files 7.3, 7.4 and 7.6.

7.9.1. Case 1. Efficiency assessment for p_2 = 1.2 kg/cm2 a

For this section consult file 7.1. Chapter 7. Case 1. Steam nozzle assessment.xlsx.

Unlike the design conditions and case 1, this case presents a strong expansion of the specific volume, but this is not fully transferred to the outflow velocity. Actual velocity (1,159 m/s) is only 5.7% lower than the ideal velocity (1,229 m/s). Instead, the specific volume experiences a strong gain: approximately 4 times of its inlet value. Flow expansion is 44% higher than geometric expansion, therefore, the nozzle is underexpanded. In this condition, oblique shock waves are generated outside the nozzle. Flow is squeezed within the nozzle and forcefully emerges with turbulence.

Table 7.8. Case 1. Expansion ratios and efficiency

p2 = 1.2 kg/cm2 a

$X_{vdesign} = 2.957$ $$X_v = \frac{V_2}{V_t} = 4.380$$	$$X_{Velocity\ design.\ Case2} = \frac{V_2}{V_t} = 1.169$$ $$X_{Velocity.\ Case1} = \frac{V_2}{V_t} = 1.736$$
$X_{f\ design.\ Case2} = 1.751$ $$X_{f\ Case1} = \frac{A_{2f}}{A_{tf}} = \frac{V_t}{V_t} \cdot \frac{V_2}{V_2} = 2.522$$	$X_{g\ Case1} = X_{g\ design.\ Case2} = 1.751$ See file 7.2 Table 7.10
$r_{A\ design.\ Case\ 2} = 1.000$ $$r_A = \frac{A_{2g}}{A_{2f}} = 0.694$$ rA < 1, underexpanded nozzle	$\varphi_{design.\ Case\ 2} = 0.970$ $e_{n\ design.\ Case\ 2} = 94.09\%$ From Formula 5.17 multiplied by 0.970: $\varphi = 0.944\ e_n = 89.03\%$

The aftermath is that the velocity coefficient was reduced from 0.970 to 0.944. The losses, which at design conditions were equal to 6% of total available energy, have now risen to 10.97% of that property. The result is that losses have increased by 82% over the design case, due to the underexpanded regime. Nonetheless this case of underexpansion is better than the overexpansion of Case 5, where total losses are equivalent to 26.7% of total available energy. See section 7.9.2.

7.9.2. Case 5. Efficiency assessment for p_2 = 5.0 kg/cm²

For this section consult file 7.5. Chapter 7. Case 5. Steam nozzle assessment.xlsx.

As discharge pressure is higher than design pressure, shock waves will be formed outside or inside the nozzle. To identify whether the shock wave is located inside or outside the nozzle, the discharge pressure must be compared with the pressure p_B of the shock wave pressure curve.

Pressure p_B value is calculated in file 7.8 with Formula 4.11 and the result is 8.25 kg/cm2 a.

Table 7.9. Flow properties for p2 = 1.2 kg/cm² a

Property	Inlet	Throat	Outlet
Mass flow kg/s	$G = \dfrac{A_t \cdot v_t}{V_t \cdot 10^4} = 0.50$		
Specific mass flow kg/s/cm²	$g_{s1} = \dfrac{G}{A_1} = 0.00$	$g_{st} = \dfrac{G}{A_t} = 0.1567$	$g_{s2} = \dfrac{G}{A_2} = 0.0895$
Sound velocity m/s	$V_{s1} = k \cdot J \cdot g \cdot R \cdot T_1 = 715$	$V_{st} = k \cdot J \cdot g \cdot R \cdot T_t = 667$	$V_{s2} = k \cdot J \cdot g \cdot R \cdot T_2 = 534$
Mach number	$M_1 = \dfrac{V_1}{V_{s1}} = 0.00$	$M_t = \dfrac{V_t}{V_{st}} = 1.000$	$M_2 = \dfrac{V_2}{V_{s2}} = 2.170$
Discharge jet power kW			$P_2 = G \cdot E_{k2} = 346$

As in Case 5 p_2 is higher than design pressure p_D but lower than p_B, geometric expansion ratio is also higher than flow expansion ratio $(X_g > X_f)$. This means that flow is overexpanded (Reminder: pressure condition of overexpansion is $p_B > p_2 > p_D$). No shock wave exists inside the nozzle but they are formed in the discharge jet.

Table 7.10. Case 5. Expansion ratios and efficiency

Formulas and results	
$X_{vdesign} = 2.957$ $X_v = \dfrac{V_2}{V_t} = 1.461$	$X_{Velocity\ design} = \dfrac{V_2}{V_t} = 1.688$ $X_{Velocity} = \dfrac{V_2}{V_t} = 1.121$
$X_{f\ design} = 1.832$ $X_f = \dfrac{A_{2f}}{A_{tf}} = \dfrac{V_t}{v_t} \cdot \dfrac{v_2}{V_2} = 1.303$	$X_{g\ design} = X_g = 1.805$
$r_{A\ design} = 1.000$ $r_A = \dfrac{A_{2g}}{A_{2f}} = 1.344$ rA > 1, overexpanded nozzle	$\varphi_{design.\ Case\ 2} = 0.970$ $e_{n\ design.\ Case\ 2} = 94.09\%$ From Formula 5.17 multiplied by 0.97: $\varphi = 0.856\ e_n = 73.27\%$

Row 1 of Table 7.100 shows that specific volume and velocity expansion actual ratios are lower than their design values. This means that nozzle performance has worsened. Velocity expansion ratio is 34% lower than design conditions and outgoing velocity is reduced in the same proportion. Furthermore, kinetic energy of outlet flow has decreased by 56%. Specific volume did not expand 3 times as in design conditions because of the higher pressure prevailing in the nozzle's divergent part.

As the nozzle is operating with outlet pressure higher than design pressure, the flow has a smaller outlet area than the geometric outlet area. The area ratio increased to 1.344, which significantly higher than 1. The nozzle is clearly overexpanded. This means that nozzle geometry is not fit to flow's shape and dimensions. This mismatching creates a turbulence that finally reduces velocity coefficient from 0.970 to 0.856. Efficiency has also suffered; under design conditions nozzle losses were 6% of the total available energy and after discharge pressure is increased, total losses grew to 45%, meaning that the total losses increased by 4.5 times.

Table 7.11. Case 5. Flow properties and jet power

Property	Inlet	Throat	Outlet	Units
Mass flow	$G = \dfrac{A_t \cdot v_t}{V_t \cdot 10^4} = 0.50$			kg/s
Specific mass flow	$g_{s1} = \dfrac{G}{A_1} = 0.0000$	$g_{st} = \dfrac{G}{A_t}$ $= 0.1567$	$g_{s2} = \dfrac{G}{A_2}$ $= 0.0895$	kg/|s. cm²|
Sound velocity	V_{s1} $= k \cdot J \cdot g \cdot R$ $\cdot T_1 = 715$	V_{st} $= k \cdot J \cdot g$ $\cdot R \cdot T_t$ $= 667$	V_{s2} $= k \cdot J \cdot g \cdot R$ $\cdot T_2 = 630$	m/s
Mach number	$M_1 = \dfrac{V_1}{V_{s1}} = 0.00$	$M_t = \dfrac{V_t}{V_{st}}$ $= 1.001$	$M_2 = \dfrac{V_2}{V_{s2}}$ $= 1.188$	NA
Discharge jet power			$P_2 = G \cdot E_{k2}$ $= 144$	kW or kJ/s

The impact on outgoing jet power of a higher discharge pressure represents 155 kW reduction, which is 51.8% of jet power under design conditions.

7.10. Cases 1 to 7. Nozzle operation sensitivity to discharge pressure

As was said before all changes to discharge pressure affect the nozzle operation. To illustrate the importance of discharge pressure variations, three significant operational parameters are calculated under different values of the outlet pressure.

The calculated parameters are:

o   Discharge jet power
o   Cross section areas of flow. Throat and outlet
o   Velocity coefficient and nozzle efficiency

The results are summarized in tables and plots of sections 7.10.1, 7.10.2 and 7.10.3. These results are consolidated in file 7.1. Chapter 7. Case 2. Steam nozzle design.xlsx.

7.10.1. Power sensitivity to discharge pressure

The results of power calculated in the above mentioned files are shown in Table 7.122 and Figure 7.6. See file 7.1. Chapter 7. Case 2. Steam nozzle design. xlsx. According to this figure the relation between power and discharge pressure is almost lineal in the pressure range of 2 to 5 kg/cm². The negative slope shows that as the discharge pressure grows the jet power comes down.

Table 7.12. Discharge jet power versus outlet pressure

Outlet pressure kg/cm2	Discharge jet power kW	% change	Expansion type
1.2	346	15.9%	Underexpansion
2.0	299	0.0%	Exact expansion
3.0	242	-19.1%	Overexpansion
4.0	185	-38.0%	Overexpansion
5.0	144	-51.7%	Overexpansion
6.0	115	-61.5%	Overexpansion

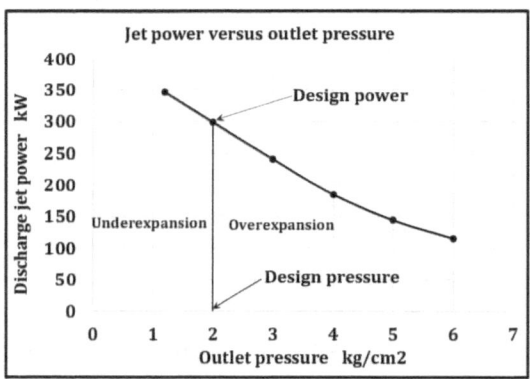

Figure 7.6. Discharge jet power versus outlet pressure

Table 7.122 shows that the underexpansion condition produces a higher power than design condition (exact expansion). Unlike underexpansion, overexpansion condition creates a scenario of lower power than design conditions.

The jet power of Case 1 is higher than power of design Case 2, because the lower discharge pressure of Case 1 produces a higher outgoing flow velocity (+6.0%) and also a higher mass flow (3%). Then the outflow kinetic energy is +12.36% higher than in Case 2 and the resulting jet power at the outlet is:

P= 1.03×1.1236×299 = 346 kW.

7.10.2. Flow cross section areas sensitivity to discharge pressure

As was discussed in section 1.8 and in Chapter 5, the cross section area of flow is a theoretical or ideal concept because any fluid (gas or steam) expands as much as possible inside the nozzle. Of course there is a limit to this expansion, which is the inner wall of the nozzle.

This physical limit imposes nozzle's dimensions and shape to the flow and the aftermath of any difference between the theoretical cross section areas and the nozzle geometry is the turbulence formation.

Minimum turbulence exists when the nozzle operates under design conditions. In practice, a common source of changes to design conditions are variations to discharge pressure. The impact of these variations are changes to the flow section areas in the divergent part.

Table 7.13 and Figure 7.7 show geometric and flow areas versus discharge pressure.

The aftermath of Figure 7.7 is that flow throat area is not affected by discharge pressure changes. However, outlet section area decreases as discharge pressure increases. As the outlet area curve has a higher slope for pressures under design pressure, it is concluded that area values sensitivity is higher in underexpanded nozzles.

Table 7.13. Geometric and flow areas versus discharge pressure

Property	p2 Discharge pressure kg/cm2 a	At Throat area cm2	A2 Outlet area cm2	% Difference Outlet flow area - Outlet geometric area
Geometric areas	NA	3.29	5.76	0.00
Flow areas	1.2	3.29	8.30	44.0%
	2.0	3.29	5.76	0.0%
	3.0	3.29	4.91	-14.8%
	4.0	3.29	4.49	-22.0%
	5.0	3.29	4.29	-25.6%
	6.0	3.29	4.17	-27.6%

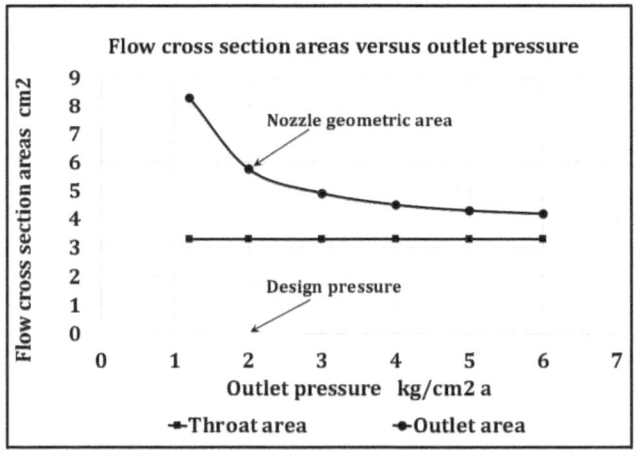

Figure 7.7. Cross section areas of flow versus outlet pressure

7.10.3. Velocity coefficient and efficiency sensitivity to discharge pressure

As discharge pressure changes, velocity coefficient also changes because the turbulence pattern is dependent of that pressure. As said before, the flow area of overexpanded nozzles is lower than underexpanded nozzles or exact expansion nozzles. Hence, the shapes difference between flow and nozzle geometry changes the turbulence pattern, which affects the velocity coefficient and Consequently, the nozzle efficiency.

Table 7.14 and Figure 7.8 display results of velocity coefficient and nozzle efficiency versus discharge pressure for the six cases under study.

Table 7.14. Velocity coefficients and expansion type versus discharge pressure

Discharge pressure kg/cm2 a	Velocity coefficient	Nozzle efficiency % of TAE	Expansion type
1.20	0.944	89.0%	Underexpanded
2.00	0.970	94.1%	Exact expansion
3.00	0.941	88.5%	Overexpanded
4.00	0.894	80.0%	Overexpanded
5.00	0.856	73.2%	Overexpanded
6.00	0.828	68.6%	Overexpanded

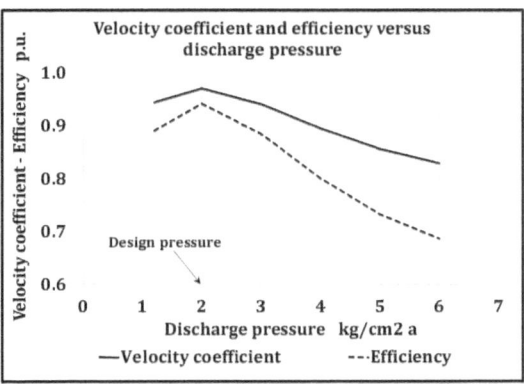

Figure 7.8. Velocity coefficient and nozzle efficiency versus discharge pressure

As expected, maximum efficiency is obtained at the design pressure. Unlike the jet power and the flow areas, efficiency does not grow in underexpansion condition.

7.11. Example of normal shock waves calculation

The discharge pressures range of Figure 7.6, to 7.8 is valid only for an underexpanded or an overexpanded nozzle, that is pressures lower than $p_B = 8.25$ kg/cm² a. In this range no shock waves are formed inside the nozzle. However, for pressures higher than p_B, shock waves are internally formed and their impact on the nozzle performance is even more harmful than the friction and turbulence consequences.

○ **Specifications of this example:**

In the nozzle designed in this Chapter, the discharge pressure is requested to be risen from 2 to 10 kg/cm² a, that is beyond the overexpanded range of pressures (2 to 8.29 kg/cm² a).

This pressure growth causes the formation of a shock wave at a distance of 1.75 cm from the nozzle inlet.

The question is to determine the shock wave impact on the nozzle performance.

The following operating data are the flow properties at the shock wave inlet. These data were drawn from file 7.8. Chapter 7. Shock wave calculation.xlsx.

○   Wave location: x = 0.951 cm
○ p_y = 4.17 kg/cm² a
○ T_y = 650 ° K
○ v_y = 0.708
○ V_y = 907 m³/kg
○ V_{sy} = 613 m/s

○ $M_z = 1.479$
○ $E_{ky} = 411$ kJ/kg

Throat properties are read in file 7.8, sheet Curves, row 30, in colored cells and red numbers. With the same colors, properties at the shock wave output are displayed in row 30. Formulas and results of this calculation are presented in Table 7.14.

Figure 7.9 represents nozzle pressure curves versus abscissa x. The shock wave is the vertical line CM. The coupling curve is ME, which smoothly links the shock wave pressure p_M with the receiving environment pressure p_E. On the curve ME, the nozzle acts as a diffuser.

In Figure 7.11, curves of Mach number before shock waves (M_y) and after a shock wave (M_z) are plotted. This graph demonstrates that the impact severity of a shock wave is reduced as the shock wave approaches the throat. That is that worst shock waves are those localized at the nozzle outlet.

Table 7.15 suggests the following comments.

○ The nozzle discharge regime changed from supersonic to subsonic.
○ The shock wave pressure rose by 134%. This sudden rise of pressure acts like a flow "plug".
○ Temperature rose by 24%. As steam is heated in the shock wave the nozzle efficiency is reduced.
○ The impact of the shock wave is noticeable in the velocities row, where the velocity reduction of the shock wave is calculated. The aftermath is that the shock wave reduces the velocity from 907 m/s to 479 m/s. Therefore, the flow kinetic energy is reduced by 72%. These figures demonstrate that a normal shock wave consequence on the nozzle performance is more important than the impact produced by the friction or turbulence.

○ Entropy increase produced by the front wave is calculated with Formula 1.6. However, the entropy change is not significant in this case; just +0.3% of entropy at the shock wave inlet. Then the actual impact on the nozzle behavior isn't the entropy increase but the kinetic energy loss of the discharge stream in the turbine.

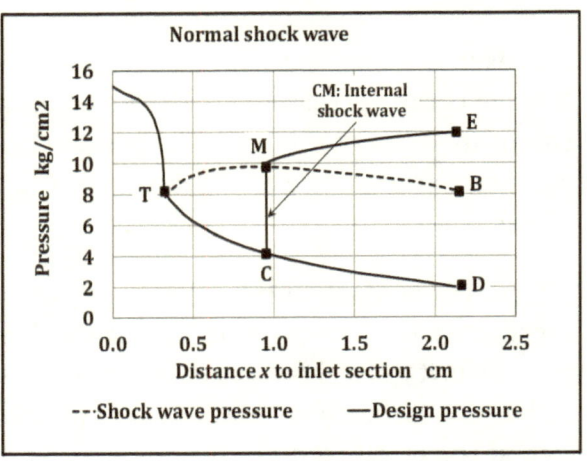

Figure 7.9. Internal shock wave formation

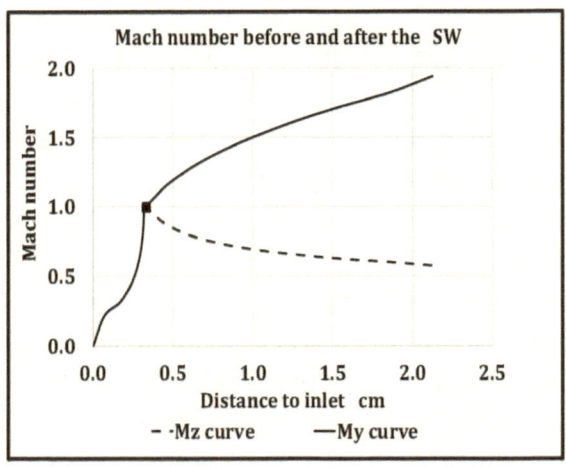

Figure 7.10. Mach number before and after a shock wave

Conclusion is that any turbine actuated by the nozzle of this example, will not properly hold an efficient operation because there exists a large kinetic energy loss caused by the shock wave. See line 8 of Table 7.15. The kinetic energy at the shock wave outlet, is now only 28% of the energy at the shock wave inlet, which represents a significant jet power reduction. Additional energy loss must be expected between the shock wave and the nozzle outlet, which will reduce even more the jet power energy.

Under these conditions, the turbine power will also be impacted by 28% in similar proportion to the energy outlet jet. This is a typical situation that must be avoided and the way to do this is to approach the actual discharge pressure as much as possible to the design pressure.

Table 7.15. Shock wave calculation

Shock wave input (Subscript y)	Formula applicable to the shock wave output	Shock wave output (Subscript z)
$M_y = 1.479$	$M_z = \sqrt[2]{\left[\dfrac{2+(k-1)\cdot M_y^2}{1-k+k\cdot M_y^2}\right]}$	$M_z = 0.702$
$p_y = 4.17$ kg/cm² a	$p_z = p_y \cdot \left[\dfrac{M_y^2 \cdot k+1}{M_z^2 \cdot k+1}\right]$	$p_z = 9.76$
Amplitude kg/cm² a	$A_z = p_z - p_y$	$A_z = 5.60$
$T_y = 650°K$	$T_z \quad T_y \cdot \left[\dfrac{2+(k-1)\cdot M}{2+(k-1)\cdot M}\right]$	$T_z = 803$

$v_y = 0.708$ m^3/kg	$v_z = v_y \cdot \dfrac{p_x \cdot T_y}{p_y \cdot T_x}$	$v_z = 0.374$
$V_{sy} = 613$ m/s	$V_{sy} = \sqrt[2]{k \cdot g \cdot J \cdot R \cdot T_y}$	$V_{sz} = 682$
$V_y = 907$ m/s	$V = M \cdot V_s$	$V_z = 479$
$E_{ky} = 411$ kJ/kg	$E_{kz} = \left(\dfrac{V_z}{V_y}\right)^2 \cdot E_{ky}$	$E_{kz} = 115$
$Sy = 8.55$ $kJ/|kg.°K|$	$\Delta S = R \cdot \ln\left(\dfrac{T_z}{T_y}\right)^{\frac{k}{k-1}} \cdot \dfrac{p_y}{p_z}$	$\Delta S = 0.03$

CHAPTER 8

DERIVATION OF FORMULAS

This chapter is not essential for the understanding and purposes of the book. It's only addressed to those professionals who are interested in the mathematical derivation of some important formulas used in the text. No new conceptual aspects are included herein.

8.1. Alternative formula for the available energy of gases

o   Temperatures of Formula 2.3 are replaced by the product of pressure times specific volume, as determined by the equation of state.

Formula 8.1. Available energy of ideal gases

$$E_a = c_p \cdot \left(T_1 - T_2 \right) = c_p \cdot \left(\frac{p_1 \cdot v_1}{J \cdot R} - \frac{p_2 \cdot v_2}{J \cdot R} \right)$$

o   From Formula 1.7 the following expression of c_p is derived.

Formula 8.2. Specific heat versus k and R

$$c_p = R \cdot \frac{k}{k-1}$$

o   Replacing c_p in Formula 8.1 the following alternate form of the available energy formula is obtained.

Formula 8.3. Available energy. Alternate formula

$$E_a = \frac{k}{J \cdot (k-1)} \cdot \left(p_1 \cdot v_1 - p_2 \cdot v_2 \right)$$

This formula allows the calculation of the total available energy in a nozzle on the basis of the initial and final thermodynamic state of a gas and its specific heat ratio k.

8.2. Conversion efficiency formula

It is demonstrated below that conversion efficiency of an ideal gas is a function of outlet and inlet temperature ratio.

Formula 8.4. Conversion efficiency for ideal gases in per unit

$$e_c = \frac{E_a}{h_1} = \frac{h_1 - h_{2ideal}}{h_1} = \frac{c_p \cdot \left(T_1 - T_{2ideal} \right)}{c_p \cdot T_1} = 1 - \frac{T_{2ideal}}{T_1}$$

o According to Formula 1.9 of isentropic relations, the temperature ratio of an ideal gas transformation is:

Formula 8.5. Temperature ratio versus pressure ratio

$$\frac{T_{2ideal}}{T_1} = \left(\frac{p_2}{p_1} \right)^{1-\frac{1}{k}}$$

o Replacing Formula 8.5 in Formula 8.4 the efficiency versus pressures ratio is given.

Formula 8.6. Conversion efficiency versus pressures ratio

$$e_c = 1 - \left(\frac{p_2}{p_1}\right)^{1-\frac{1}{k}}$$

This formula demonstrates that as nozzle pressure ratio increases, conversion efficiency decreases. That's why C nozzles have lower conversion efficiency than CD nozzles.

8.3. Saint Venant - Wantzel equation

This equation was originally derived to calculate the outgoing flow velocity from an orifice. At the end of this section it will be specifically implemented to calculate the throat velocity in CD nozzles and the discharge velocity in C nozzles.

To derive this equation, the following substitutions are made in Formula 1.1 of energy conservation:

○   Enthalpies are replaced by Formula 1.10.

Formula 8.7. Equation of energy conservation

$$u_1 + \frac{p_1 \cdot v_1}{J} + \left(\frac{V_1^2}{1.g.J}\right) = u_x + \frac{p_x \cdot v_x}{J} + \left(\frac{V_x^2}{1.g.J}\right)$$

○   Internal energies are replaced by Formula 8.8 and Formula8.9

Formula 8.8. Input internal energy

$$u_1 = c_v \cdot T_1 = \left(c_p - R\right) \cdot T_1$$

Formula 8.9. Internal energy

$$u_x = c_v \cdot T_x = \left(c_p - R\right) \cdot T_x$$

o T_1 and T_x are replaced by the following expressions derived
   from the equation of state:

Formula 8.10. Input temperature

$$T_1 = \frac{p_1 \cdot v_1}{J.R}$$

Formula 8.11. Temperature at any location

$$T_x = \frac{p_x \cdot v_x}{J.R}$$

o The specific volume v_x is replaced by Formula 8.12 derived
   from Formula 1.9 of isentropic relations.

Formula 8.12. Specific volume at any location

$$v_x = v_1 \cdot \left(\frac{p_1}{p_x}\right)^{\frac{1}{k}} = \frac{v_1}{\left(r_{px}\right)^{\frac{1}{k}}}$$

o The introduction of Formulas 8.8, 8.9, 8.10, 8.11 and 8.12 in
   Formula 8.7, leads to the Saint Venant - Wantzel formula of
   velocity (see Formula 8.13).

Formula 8.13. Saint Venant–Wantzel equation

$$V_x\left(r_{px}\right) = \sqrt[2]{\frac{2 \cdot g \cdot k}{k-1}\left[1-\left(r_{po}\right)^{\frac{k-1}{k}}\right]} \times \sqrt[2]{P_1 \cdot v_1}$$

Where r_{po} is the orifice pressure ratio p_o/p_1.

○   CD nozzle velocity at the throat:

Formula 8.14. Throat velocity of a CD nozzle

$$V_t\left(r_{pt}\right) = \sqrt[2]{\frac{2 \cdot g \cdot k}{k-1}\left[1-\left(r_{pt}\right)^{\frac{k-1}{k}}\right]} \times \sqrt[2]{P_1 \cdot v_1}$$

Where: $r_{pt} = p_t/p_1$

○   C nozzle discharge velocity

Formula 8.15. Discharge velocity of a C nozzle

$$V_2\left(r_{pn}\right) = \sqrt[2]{\frac{2 \cdot g \cdot k}{k-1}\left[1-\left(r_{pn}\right)^{\frac{k-1}{k}}\right]} \times \sqrt[2]{P_1 \cdot v_1}$$

Where r_{pn} is the nozzle pressure ratio: $r_{pn} = p_2/p_1$

8.4. Critical ratios formulas

○   By equating Formula 3.4 of Saint Venant - Wantzel with Formula 3.6 of the speed of sound at the throat, the following equation is obtained.

Formula 8.16. Speed of sound and Saint
Venant - Wantzel identity

$$\sqrt[2]{k \cdot g \cdot J \cdot R \cdot T_t} = \sqrt[2]{\frac{2.k.g}{k-1} \cdot \left[1 - \left(r_{pt}\right)^{\frac{k-1}{k}}\right] \cdot p_1 \cdot v_1}$$

○ As this equation meets the critical condition $(V_t = V_{sound})$, the
following formulas are replaced in Formula 8.16

Formula 8.17. Pressure ratio at the throat. Choked flow

$$r_{pt} = r_{pc} = \frac{p_t}{p_1}$$

Formula 8.18. Equation of state at the throat. Choked flow

$$p_{tc} \cdot v_{tc} = J \cdot R \cdot T_{tc}$$

○ These substitutions return the following expression.

Formula 8.19. Formula of $(p_{tc} \cdot v_{tc})$

$$p_{tc} \cdot v_{tc} = \frac{2.k.g}{k-1} \cdot \left[1 - \left(r_{pc}\right)^{\frac{k-1}{k}}\right] \cdot p_1 \cdot v_1$$

This equation relates the fluid thermodynamic state at the
throat, with the thermodynamic state at the inlet.

○ The following isentropic relation is replaced in Formula 8.19.

Formula 8.20. Specific volumes ratio

$$\frac{V_1}{V_t} = \left(\frac{p_t}{p_1}\right)^{\frac{1}{k}} = r_{pc}^{\frac{1}{k}}$$

o All these replacements produce a first version of the critical pressure ratio.

Formula 8.21. Critical pressure ratio versus r_{pc}

$$r_{pc} = \frac{p_{tc}}{p_1} = \frac{2.k.g}{k-1} \cdot \left[1 - \left(r_{pc}\right)^{\frac{k-1}{k}}\right] \cdot r_{pc}^{\frac{1}{k}}$$

Where p_{tc} is the critical pressure at the throat.

o By clearing r_{pc} from Formula 8.21 the critical pressure ratio is obtained.

Formula 8.22. Critical pressure ratio formula

$$r_{pc} = \frac{p_{tc}}{p_1} = \left(\frac{2}{k+1}\right)^{\frac{k}{k-1}}$$

o Temperature and specific volume critical ratios are derived from the following isentropic relations.

Formula 8.23. Critical temperature ratio formula

$$\frac{T_{tc}}{T_1} = \left(\frac{p_{tc}}{p_1}\right)^{\frac{k-1}{k}}$$

Formula 8.24. Critical specific volume ratio formula

$$\frac{V_{tc}}{V_1} = \left(\frac{P_1}{P_{tc}}\right)^{\frac{1}{k}}$$

o   Introducing Formula 8.22 in Formulas 8.23 and 8.24, the critical temperature and specific volume ratios formulas are obtained.

Formula 8.25. Critical temperature ratio

$$r_{tc} = \frac{T_{tc}}{T_1} = \frac{2}{k+1}$$

Formula 8.26. Critical specific volume ratio

$$r_{vc} = \frac{V_{tc}}{V_1} = \left(\frac{2}{k+1}\right)^{\frac{1}{k-1}}$$

8.5. Mass flow versus nozzle pressure ratio

o   The mass flow at any section x of the nozzle is given by the principle of continuity, which states that mass flow is the same at any cross section of the nozzle. Therefore, mass flow canbe calculated at any convenient cross section. Usually it's adopted the throat section in CD nozzles or discharge section in C nozzles.

Formula 8.27. Mass flow general formula

$$G = \frac{A_o \cdot V_o}{V_o}$$

○ Where V_o is replaced by the Saint Venant - Wantzel equation (Formula 8.13).

Formula 8.28. Saint Venant–Wantzel equation

$$V_o = \sqrt[2]{\frac{2 \cdot g \cdot k}{k-1} \left[1 - \left(r_{po}\right)^{\frac{k-1}{k}}\right]} \times \sqrt[2]{P_1 \cdot V_1}$$

○ The specific volume v_o is replaced by:

Formula 8.29. Specific volume v_o

$$V_o = \frac{V_1}{r_{po}^{\frac{1}{k}}}$$

Formula 8.30. Pressure ratio formula at the throat (CD nozzles) or discharge (C nozzles)

$$r_{po} = \left(\frac{P_o}{P_1}\right)$$

Where r_{po} has been defined in the group of Formulas 3.5.

○ After these two replacements are done, the following formula of mass flow is given, which is valid for the throat of CD type nozzles or for the discharge of C type nozzles.

Formula 8.31. Mass flow formula

$$G\left(r_{po}\right) = \sqrt[2]{\frac{2.k.g}{k-1} \cdot \left[\left(r_{po}\right)^{\frac{2}{k}} - \left(r_{po}\right)^{\frac{k+1}{k}}\right]} \cdot A_o \cdot \frac{P_1}{V_1}$$

○ The curve of $G(r_{po})$ versus the pressure ratios is plotted in Figure 3.3. This curve has two zeros; the first is at $r_{po} = 0$ and the second is at $r_{po} = 1$. The general curve shape looks like a semicircle, with a maximum at the point where the flow throttling starts. This maximum occurs at $r_{po} = r_{pc}$. Therefore, by replacing in Formula 8.31 A_o by A_t and r_{po} by Formula 3.7, the maximum mass flow formula is obtained.

Formula 8.32. Maximum mass flow

$$G_{max} = \left[\sqrt[2]{g \cdot k \cdot \left(\frac{2}{k+1} \right)^{\frac{k+1}{k-1}}} \right] \cdot \sqrt[2]{\frac{p_1}{v_1}} \cdot A_t$$

8.6. Thermodynamic properties versus Mach number

In an ideal nozzle the friction loss of energy q_f and the inlet fluid velocity V_1 are negligible. Consequently, the principle of energy conservation formula is as follows.

Formula 8.33. Input enthalpy

$$h_1 = h_x + \left(\frac{V_x^2}{2.g.J} \right)$$

Where enthalpies are substituted by the general enthalpy formula $c_p.T$. By doing this the T_1/T_x ratio, is cleared and the result is the following formula.

Formula 8.34. Temperature ratio at any location and inlet

$$\frac{T_1}{T_x} = 1 + \frac{V_x^2 \cdot (k-1)}{2 \cdot k \cdot g \cdot J \cdot R \cdot T_x}$$

Where T_x is the temperature at any distance x from the inlet,

According to Formula 3.6 the product $(k \cdot g \cdot J \cdot R \cdot T_x)$ equals the square of the sound velocity V_{sx}. Substituting this expression in Formula 8.34, the T_1/T_x ratio formula results:

Formula 8.35. Temperature ratio versus Mach number

$$\frac{T_1}{T_x} = 1 + \frac{k-1}{2} . M_x^2$$

Analogous ratios are derived for pressure and specific volumes by using the isentropic relations of Formula 1.9. The resulting expressions are Formulas 8.36 and 8.37.

Formula 8.36. Pressure ratio versus temperature ratio

$$\frac{P_x}{P_1} = \left(\frac{T_x}{T_1} \right)^{\frac{k}{k-1}}$$

Formula 8.37. Specific volume ratio versus temperature ratio

$$\frac{V_x}{V_1} = \left(\frac{T_1}{T_x} \right)^{\frac{1}{k-1}}$$

Introducing Formula 8.35 in Formulas 8.36 and 8.37 the temperature and specific ratios versus Mach number are obtained.

Formulas 8.38. State properties ratios versus Mach number

$$\frac{T_x}{T_1} = \frac{2}{2+(k-1)\cdot M_x^2} \qquad \frac{p_x}{p_1} = \left[\frac{2}{2+(k-1)\cdot M_x^2}\right]^{\frac{k}{k-1}}$$

$$\frac{V_x}{V_1} = \left(1+\frac{k-1}{2}.M_x^2\right)^{\frac{1}{k-1}}$$

This group of formulas is used to link the state properties at the inlet and the throat as a function of the Mach number at the throat, as follows.

Formulas 8.39. State properties ratios. Inlet and throat

$$\frac{T_t}{T_1} = \frac{2}{2+(k-1)\cdot M_t^2} \qquad \frac{p_t}{p_1} = \left[\frac{2}{2+(k-1)\cdot M_t^2}\right]^{\frac{k}{k-1}}$$

$$\frac{V_t}{V_1} = \left(1+\frac{k-1}{2}.M_t^2\right)^{\frac{1}{k-1}}$$

By dividing each formula of group of Formulas 8.38 by the corresponding formula of group of Formulas 8.39, a general relationship between the throat properties and properties at any section of the nozzle is obtained.

Formulas 8.40. Properties of state ratios versus Mach number

$$\frac{T_x}{T_t} = \frac{2+(k-1)\cdot M_t^2}{2+(k-1)\cdot M_x^2}$$

$$\frac{p_x}{p_t} = \left[\frac{2+(k-1)\cdot M_t^2}{2+(k-1)\cdot M_x^2}\right]^{\frac{k}{k-1}}$$

$$\frac{v_x}{v_t} = \left[\frac{2+(k-1)\cdot M_x^2}{2+(k-1)\cdot M_t^2}\right]^{\frac{1}{k-1}}$$

If the nozzle is CD type the Mach number M_t at the throat equals 1, therefore, the group of Formulas 8.40 is converted into the following group.

Formulas 8.41. Properties of state ratios
versus Mach number. Choked flow

$$\frac{T_x}{T_t} = \frac{k+1}{2+(k-1)\cdot M_x^2}$$

$$\frac{p_x}{p_t} = \left[\frac{k+1}{2+(k-1)\cdot M_x^2}\right]^{\frac{k}{k-1}}$$

$$\frac{v_x}{v_t} = \left[\frac{2+(k-1)\cdot M_x^2}{k+1}\right]^{\frac{1}{k-1}}$$

The geometric profile of the nozzle is designed with these relationships. See section 8.7.

8.7. Cross section area versus pressure ratio r_{px}

The formula to be derived is only valid for ideal gas and is obtained by replacing the following isentropic relations in Formula 5.3.

Formula 8.42. Temperature ratio in Formula 5.3

$$\frac{T_x}{T_t} = \left(\frac{p_x}{p_t}\right)^{\frac{k-1}{k}} = \left(\frac{p_x/p_1}{p_t/p_1}\right)^{\frac{k-1}{k}} = \left(\frac{r_{px}}{r_{pt}}\right)^{\frac{k-1}{k}}$$

Formula 8.43. Pressure ratio in Formula 5.3

$$\frac{p_t}{p_x} = \left(\frac{p_t/p_1}{p_x/p_1}\right) = \frac{r_{p1}}{r_{px}}$$

Formula 8.44. Enthalpies in Formula 5.3

$$h_1 = c_p \cdot T_1 \qquad h_t = c_p \cdot T_t \qquad h_x = c_p \cdot T_x$$

By replacing Formulas 8.42, 8.43 and 8.44 in Formula 5.3, the expression of cross section area versus pressure ratios is returned.

Formula 8.45. Cross section areas ratio
versus pressure ratio for gas nozzles

$$\frac{A_x}{A_t} = \left(\frac{r_{pt}}{r_{px}}\right)^{\frac{1}{k}} \cdot \sqrt[2]{\frac{1-\left(r_{pt}\right)^{\left(1-\frac{1}{k}\right)}}{1-\left(r_{px}\right)^{\left(1-\frac{1}{k}\right)}}}$$

This formula is named Formula 5.4 in Chapter 5.

8.8. Cross section area versus Mach number

Mass flow formula is used to determine the flow shape and dimensions.

According to the principle of continuity, mass flow at any section x is given by Formula 8.46.

Formula 8.46. Mass flow

$$G_x = \frac{A_x \cdot V_x}{v_x}$$

Mass flow formula at the throat is:

Formula 8.47. Mass flow at the throat

$$G_t = \frac{A_t \cdot V_t}{v_t}$$

According to the continuity principle, mass flow G_x equals G_t and from this equality derives the flow area formula.

Formula 8.48. Cross section area at any location

$$A_x = \frac{v_x}{v_t} \cdot \frac{V_t}{V_x} \cdot A_t$$

Alternatively, a formula of A_x with M_x as independent variable is derived, by doing the following replacements in Formula 8.48 as follows.

○ Replace the v_x/v_t specific volume ratio with the appropriate expression of Formulas 8.40
○ Replace the V_t/V_x ratio with the following expression:

Formula 8.49. Velocity ratio versus Mach
number ratio at any location and throat

$$\frac{V_t}{V_x} = \frac{V_{st} \cdot M_t}{V_{sx} \cdot M_x} = \frac{\sqrt[2]{k \cdot g \cdot J \cdot R \cdot T_t}}{\sqrt[2]{k \cdot g \cdot J \cdot R \cdot T_x}} \cdot \frac{M_t}{M_x} = \sqrt[2]{\frac{T_t}{T_x}} \cdot \frac{M_t}{M_x}$$

Where T_t/T_x ratio is replaced by the analogous expression of Formulas 8.41.

The results of these replacements is an alternative formula for the area at any point x of the nozzle, versus the Mach number M_x at that location.

Formula 8.50. Cross section area

$$A_x = \left[\frac{2 + (k-1) \cdot M_x^2}{2 + (k-1) \cdot M_t^2} \right]^{\frac{1}{k-1}} \cdot \sqrt[2]{\frac{2 + (k-1) \cdot M_x^2}{2 + (k-1) \cdot M_t^2}} \cdot \frac{M_t}{M_x} \cdot A_t$$

This formula can be simplified and this operation returns the following expression of A_x.

Formula 8.51. Cross section area versus Mach number

$$A_x = \left[\frac{2 + (k-1) \cdot M_x^2}{2 + (k-1) \cdot M_t^2} \right]^{\frac{k+1}{2 \cdot (k-1)}} \cdot \frac{M_t}{M_x} \cdot A_t$$

For CD nozzles, which usually operate under critical conditions, the Mach number in the throat is unity. Therefore, Formula 8.51 is converted into the following simpler expression.

Formula 8.52. Nozzle area at any point x

$$A_x = \left[\frac{2 + (k-1) \cdot M_x^2}{k+1} \right]^{\frac{k+1}{2 \cdot (k-1)}} \cdot \frac{A_t}{M_x}$$

The conical nozzle diameter D_x under critical conditions is derived from Formula 8.52 and the result is the following formula.

Formula 8.53. Nozzle diameter at any section x

$$D_x = \left[\frac{2 + (k-1) \cdot M_x^2}{k+1} \right]^{\frac{k+1}{4 \cdot (k-1)}} \cdot \frac{1}{\sqrt[2]{M_x}} \cdot D_t$$

BIBLIOGRAPHY

• **Chapter 1**

[1]   Wikipedia, "History of the Steam Engine," 5 October 2015. [Online]. Available: https://en.wikipedia.org/wiki/History_of_the_steam_engine. [Accessed June 2014].

[2]   Wikipedia, "History of Heat," 18 July 2015. [Online]. Available: https://en.wikipedia.org/wiki/History_of_heat. [Accessed August 2015].

[3]   S. Carnot, Reflections on the Motive Power of Fire, Paris, France: Chez Bachelier, Libraire, 1824.

• **Chapter 2**

[4]   USA Navy, Energy Analysis of Naval Machinery, Annapolis, Maryland, USA: US Naval Institute, 1940.

[5]   Y. Cengels, Thermodynamics. An engineering approach. Third Edition, WCB/McGraw-Hill, 1998.

• **Chapter 3**

[6]   NASA Glen Research Center, "Nozzles," May 2015. [Online]. Available: https://www.grc.nasa.gov/www/k-12/airplane/nozzle.html. [Accessed 2015].

[7]   K. Rolle, Thermodynamics and Heat Power, Prentice Hall, 2005.

[8]   K. V. Wong, Thermodynamics for Engineers, Boca Raton, FL, USA: CRC Press. Taylor and Francis Group, 2012.

[9] A. Estrada, Termodinámica Técnica, Buenos Aires, Argentina: Librería y Editorial Alsina, 1955.

[10] M. Somerton, Thermodynamics for Engineers, New York, NY, USA: McGraw-Hill, 1993.

[11] B. Sonntag, Thermodynamic and Transport Properties, New York, NY, USA: John Wiley and Sons, Inc., 1997.

• Chapter 4

[12] J. Welty, Fundamentals of Momentum, Heat and Mass Transfer, New York, NY USA: John Wiley and Sons, Inc, 2001.

[13] R. D. Zucker, Fundamentals of Gas Dynamics, Chesterland, OH, USA: Matrix Publishers Inc., 1977.

[14] B. a. Hughes, Fluid Dynamics, New York, NY, USA: McGraw-Hill Book Company, Inc, 1967.

[15] F. M. White, Fluid Mechanics, McGraw-Hill Inc. USA, 1979.

[16] I. H. Shames, Mechanics of Fluids, New York, NY, USA: The McGraw-Hill Companies, Inc, Copyright at 2003. First Editoin 1962.

[17] Dickson, Fluid Mechanics and Thermodynamics of Turbomachinery, New York: Elsevier, 2013.

[18] R. Granger, Fluid Mechanics, 1994: Dover.

[19] J. Skorpik, "Flow of gases and Steam Through Nozzles," March 2014. [Online]. Available: http://www.transformacni-technologie.cz. [Accessed 2014].

[20] B. Lakshminarayana, Fluid Dynamics and Heat Transfer of Turbomachinery, Hoboken, NJ, USA: John Wiley and Sons, Inc., 1996.

• **Chapter 5**

[21] R. Patel, Elements of Heat Engines. Volumes I, II and III, Baroda, India: Acharya Publications, 1st Edition 1963 - 18th Edition 1997.

[22] N. M. Mahmood, "Diyala Journal of Engineering Sciences. Simulation of Back Pressure Effect on Behavior of Convergent Divergent Nozzle," March 2013. [Online]. Available: http://www.iasj.net/iasj?func=fulltext&aId=71720. [Accessed 2015].

[23] P. José Luis Rodríguez, "Ondas de Choques en Toberas (Standing Shock Waves in Nozzles)," March 2004. [Online]. Available: http://www.unet.edu.ve/~jlrodriguezp/ochoque.pdf. [Accessed 2015].

[24] Korakianitis., The Design of High Efficiency Turbomachinery, Massachusets: Massachusets Institute of Technology, 2014.

• **Chapter 6**

[25] M. P. Boyce, Gas Turbine Engineering Handbook, Waltham, MA, USA: Elsevier, 2012.

[26] P. Schilke. GE Energy, "Advanced Gas Turbine Materials and Coatings," 2004. [Online]. [Accessed 2014].

• **Chapter 7**

[27] E. F. Church, Steam Turbines, New York and London: McGraw-Hill Bokk Company, Inc, 1935.

[28] P. Shlyakhin, Steam Turbines. Theory and Design, Honolulu, Hawaii, USA: University Press of the Pacific, Copyright at 2005.

[29] A. Stodola, Steam Turbines, New York, NY, USA: D. Van Nostrand Company, 1905.

[30] G. Belluzzo, Le Turbine a Vapore Ed a Gas, Milano: Editore Libraio Della Real Casa, 1905.

[31] H. Dubbel, Handbook of Mechanical Engineering, London, UK: Springer Verlag, Wolfgang Beitz (Editor), B.J. Davies (Editor), Karl-Heinz Küttner (Editor), 1994.

[32] Baumeister, Standard Handbook for Mechanical Engineers, New York, NY, USA: McGraw-Hill, 2007.

List of Figures

Figure 1.1. Convergent divergent nozzle 7

Figure 1.2. Convergent nozzle 7

Figure 1.3. Diffuser 7

Figure 1.4. Nozzle profile and pressure distribution.................... 17

Figure 1.5. Steam temperature and specific volume
distribution 17

Figure 1.6. Steam enthalpies and energies.................................... 18

Figure 1.7. Steam and sound velocities and Mach number
distribution 18

Figure 1.8. Specific mass flow and sections area curves........... 21

Figure 1.9. Turbulent flow inside a CD nozzle 23

Figure 1.10. Velocity and mass flow behavior versus the
discharge pressure 31

Figure 2.1. Conversion efficiency of inlet enthalpy
h_1. Steam curve is an approximation 35

Figure 2.2. Nozzle expansion in the Mollier diagram................. 40

Figure 3.1. Saint Venant - Wantzel equation. Subscripts
interpretation 44

Figure 3.2. Flow velocity curve versus pressure ratio............... 51

Figure 3.3. Flow mass curve versus pressure ratio 53

Figure 4.1. Moody diagram...66

Figure 4.2. Velocity coefficient versus L/D ratio
and friction factor. Darcy – Weisbach formula 68

Figure 4.3. Reynolds number distribution for different
discharge pressures 75

Figure 4.4. Reynolds number versus pressure
and temperature.. 76

Figure 4.5. Shock wave coefficients. State variables...................81

Figure 4.6. Shock wave coefficients. Flow properties 81

Figure 4.7. Shock wave regions...85

Figure 4.8. Shock wave amplitude 85

Figure 4.9. Internal shock wave..86

Figure 4.10. Normal shock wave at the outlet..............................86

Figure 5.1. Areas ratio versus pressure ratio 96

Figure 5.2. Nozzle profile and sections area curve versus
pressure ratio...98

Figure 5.3. Areas ratio versus Mach number 100

Figure 5.4. Flow shape and turbulence inside a conical flow. 103

Figure 5.5. Identification curves of nozzle expansion.............107

Figure 5.6. Velocity coefficient. Steam nozzles. US Naval
Institute chart...110

Figure 5.7. Nozzle velocity coefficient for steam nozzles........ 111

Figure 5.8. Steinmetz curve 112

Figure 5.9. Steinmetz and US Naval Institute curves
        with same maximum velocity coefficient............. 113

Figure 5.10. Nozzle design flow chart 115

Figure 5.11. Example of conical nozzle dimensions versus
        velocity coefficient.. 123

Figure 5.12. Thermodynamic properties versus pressure 127

Figure 5.13. Pressure curve and nozzle profile.......................... 127

Figure 5.14. Temperature and specific volume curves 128

Figure 5.15. Enthalpy and velocity curves 128

Figure 6.1. Nozzle profile and pressure curve.............................147

Figure 6.2. Enthalpies, kinetic energy and total energy
        versus distance to inlet.....................................147

Figure 6.3. Temperature and specific volume curves 148

Figure 6.4. Flow velocity, sound velocity and Mach
        number curves... 148

Figure 7.1. Nozzle assessment flow chart 151

Figure 7.2. Nozzle profile and pressure curve.............................161

Figure 7.3. Energy curves...161

Figure 7.4. Temperature and specific volume curves 161

Figure 7.5. Velocities and Mach number curves............................ 162

Figure 7.6. Discharge jet power versus outlet pressure 168

Figure 7.7. Cross section areas of flow versus outlet
pressure..170

Figure 7.8. Velocity coefficient and nozzle efficiency
versus discharge pressure 171

Figure 7.9. Internal shock wave formation 174

Figure 7.10. Mach number before and after a shock wave.......174

List of Uploaded Files

1.1. Figures 1.4, 1.5, 1.6, 1.7 and 1.8. Performance curves.xlsx

1.2. Figure 1.9. Velocity and mass flow versus p_1.xlsx

2.1. Figure 2.1. Conversion efficiency of h_1.xlsx

3.1. Figures 3.1 and 3.2. Mass flow and velocity versus pressure ratio.xlsx

4.1. Figure 4.1. Moody diagram.jpg

4.2. Table 4.1. Efficiency sensitivity to roughness.xlsx

4.3. Figure 4.2. Velocity coefficient and Darcy formula.xlsx

4.4. Figures 4.3 and 4.4. Reynolds assessment.xlsx

4.5. Figures 4.5 and 4.6. Shock waves coefficients.xlsx

4.6. Tables 4.2 and 4.3. Shock wave calculation.xlsx

5.1. Figures 5.1 and 5.3. Area ratios tables.xlsx

5.2. Figure 5.2. CD nozzle profile versus r_p.xlsx

5.3. Figure 5.4. Flow shape in a CD nozzle.ppt

5.4. Figures 5.5. Expansion identification.xlsx

5.5. Figures 5.6 and 5.7. Velocity coefficient and efficiency charts.xlsx

5.6. Figure 5.8. Geometric design. Flow chart.ppt

5.7. Figure 5.9. Nozzle dimensions sensitivity.xlsx

5.8. Figures 5.10 to 5.14. Example of a CD nozzle geometric design.xlsx

6.1. Chapter 6. Gas nozzle project.xlsx

7.1. Chapter 7. Case 2. Steam nozzle design.

7.2. Chapter 7. Case 1. Steam nozzle assessment.xlsx

7.3. Chapter 7. Case 3. Steam nozzle assessment.xlsx

7.4. Chapter 7. Case 4. Steam nozzle assessment.xlsx

7.5. Chapter 7. Case 5. Steam nozzle assessment.xlsx

7.6. Chapter 7. Case 6. Steam nozzle assessment.xlsx

7.8. Chapter 7. Shock wave calculation.xlsx

INDEX

A

abscissa x 3, 6

adiabatic 13

aircraft engines 5

angle

    of convergence 6

    of divergence 6

B

blade wheels 6

C

CD nozzle 7

    inlet velocity 19

choked

    pressures condition 28

chokes 27

circulation time 19

C nozzle 7

combustion chambers 8

combustion gases 8

compressor 8

condenser 41

condition curve 41

constricted section 19

control volume 9

convergent nozzle 6, 7

conversion efficiency conversion efficiency 34

cover sheath 7

critical

    condition 27

    conditions 45

    pressure ratio formula 46

    ratio calculation 29

    ratios 28

    ratios formulas 28, 29

    regime 28

    specific volume ratio formula 46

    state variables 28

    temperature ratio formula 46

    thermodynamic properties 20

    velocity 27

critical conditions. 20

critical pressure 20

critical properties 20

critical ratios 45, 46

cross section 6, 12, 21

    area calculation 93

    area ratio versus pressure ratio 94

    areas ratio Ax/At 94

    areas ratio versus Mach number 99

    ratio versus p, T and h 94

D

Darcy – Weisbach formula 62

De Laval 20

derivation of formulas

    available energy of gases 177

    conversion efficiency 178

    critical ratios formulas 181

    cross section area versus Mach 191

    mass flow 184

    properties versus Mach number 186

    Saint Venant - Wantzel equation 179

diffuser 7

diffusers 5, 6, 9

discharge pressure 20

discharge stream 20

discontinuity lamina 25

discontinuity lamina thickness 26

distribution 3

divergent part 20

E

efficiency

    conversion 32

    conversion of input enthalpy 35

elastic behavior of gas molecules 27

elasticity 27

energy

    available 16, 32

    available in a gas nozzle 33

    conversion efficiency 34

    Darcy - Weisbach formula 63

    exchange 16

    forms of 9

    inflow 9

    internal 9

    kinetic 9

    kinetic loss 63

    outflow 9

    potential 9

    pressure 9

    TAE 32

TAE formula 33

thermal 16

total available 32

enthalpy

actual discharge 11, 37, 39

gain by the fluid 36

general formula 15

ideal at the outlet 10

ideal gas 15

increase 36

saturated steam 34

superheated steam 33

wet steam 34

entropy 11, 13

growth 13

increase 13, 41

equation of state 8, 11, 14

expansion 16

geometric 56

in a nozzle 39

in Mollier diagram 40

specific volume 39

specific volume 56

velocity 56

expansion ratios 104

actual 105

calculation procedure 123

exact expansion 106

expansion curves
    identification 106

flow 105

geometric 105

optimum relation 105

overexpansion 106

underexpansion 106

velocity coefficient 108, 111

external environment 9

F

flow

behavior 24

choked 26, 45

choking 26

complete turbulent 24

complete turbulent
    region 66

discharge velocity 42

fractioned in turbine
    blades wheel 60

friction losses 8

laminar 65

one-dimensional 3, 17

outgoing 20

outlet kinetic energy 67

plume shape 20

regime compressible
and viscous 3

scattered 20

strangled 26

supersonic 20, 24

transition region 65

tube shape 20

turbulence 6

turbulent 23, 24

turbulent region 66

velocity 12, 26, 45

velocity behavior 19

velocity growth 26

flow shape

cross section area
formula 104

minimum turbulence 102

optimum design
condition 102

optimum geometric
design 104

flow velocity 42, 43

fluid

enthalpy 32

gravity 30

vein 26

fluid enthalpy 32

friction 3

absence, ideal 32

factor 63

losses 16, 40

released heat 67

released heat 9

friction factor 23, 24, 62, 67

Colebrook formula 65

Moody diagram 65

Moody formula 65

versus L/D ratio 68

G

gas constant

specific 11

universal 11

gas constant and specific
heats 14

gas nozzle project 133

calculation organization 133

critical properties and
type 137

Data and Physical
Constants 134

energetic properties 139

expansion ratios 145

flow properties 141

geometric design 143

performance curves 147

thermodynamic properties 138

gas turbines 8

geometric design

velocity coefficient and dimensions 122

geometric expansion 58

goose egg shape. 7

gravity acceleration 10

GT air compressor 8

I

ideal thermodynamic transformations 13

industrial machines 5

inject combustion gas 8

internal energy 15

isentropic 13

condition 14

relations 13, 14

J

James Watt 4

jet shape 20

K

kinetic energy 6

curves 18

L

lifting water 4

M

mass flow 42, 45

choked flow 51

constant aG formula 54

curve 30

curve versus nozzle ratio 53

density 21

detailed formula 54

maximum value 54

specific 21

versus nozzle pressure ratio 52

mechanical equivalent 10

molecular disorder 13

molecular weight 12

Mollier diagram 39, 40

momentum 78

N

naval propulsion 5

neck valve 104

net heat exchanged 9

net mechanical work 9

normal shock wave 25

nozzle

average diameter 63

body 36

convergent 5

convergent divergent 5

convergent part 19

divergent part 19

inlet pressure 30

inner wall 10

performance 51

performance curves.
    Example 124

pressure ratio 28

total length 63

types 5

nozzle efficiency 24

O

oil and gas fields 27

one-dimensional 3

optimum design 105

Osborne Reynolds 27

Outflow energy 9

P

physical laws 8

pipe 23

power generation 5

pressure

    design 24

    discharge 24

    discharge jet 26

    discharge regulation 24

distribution 20

energy 8

external 26

ratio 45

upstream 27

vessel 44

principle

    conservation of energy 8, 9

    conservation of the fluid's
        momentum 9

    mass continuity 9

    of continuity 93

properties

    state 12

R

random movement 13

receiving environment 25, 82

reheat 37

reversible 13

Reynolds number 23, 62

    equivalent diameter 72

    formula 64

rugosity 62, 63

    absolute 63

    relative 63

S

Sadi Carnot 4

Saint Venant Jean Claude 44

Saint Venant - Wantzel
    equation 44

  caveat 45

sheath diffuser 8

shock wave 25, 26, 89

  amplitude plot 84

  coefficients 82

  Curves. Point A 83

  Curves. Point B 83

  Curves. Point D 83

  Curves. Point T 83

  design discharge
     pressure 84

  discontinuity 25

  entropy 25

  entropy increase 25

  exit kinetic energy 81

  exit Mach number 78

  exit pressure formula 79

  exit specific volume 79

  exit temperature formula 79

  exit velocity 80

  extreme curves points 83

  formation 25

  in CD nozzle 24

  inlet 26

  Mach number at point B 84

Mach number at point D 84

  normal 25

  normal internal or oblique
    external 88

  overexpanded 87

  plug 25

  pressure at point B 83

  pressure behavior
    summary 83

  principle of continuity 93

  receiving environment
    pressure 87

  subscripts nomenclature 78

  thickness 26

  turbine 91

  turbulent regime 88

  underexpanded 87

  zone of instability 25

sonic flow 20

spatial coordinate 3

specific

  gravity 12

  heat 43, 46

  heats ratio 14, 27

  mass flow 20

  volume 12

  volume expansion 57

speed of sound 26, 27, 45, 49, 61, 80

    formula 45

state properties 15, 25

steam

    saturated 12

    superheated 12, 15, 34, 47

    title 39

    transformations in Mollier diagram 40

steam nozzle 39, 53

steam nozzle project

    efficiency assessment 162

    geometric design 158

    normal shock wave calculation 172

    performance curves 160

    specifications 151

steam turbines 8

straight section 6

stream shape, 20

subscripts

    nomenclature 6

surface facilities 27

system of units xvii

T

TAE

    input velocity inclusion 39

thermal energy 15

thermodynamic properties 14

thermodynamic transformation 25

Thomas Savery 4

throat

    pressure 27

thrust 8

turbine

    blades 20

    operation 25

    power 42

    power formula 91

    rotor 20

    shaft 8

turbulence 20, 22, 106

    severity 24

V

velocities

    canceled 27

    in opposition 27

velocity

    actual discharge 43

    constant of general formula 49

    general formula at the throat 49

    ideal discharge 42, 43

velocity coefficient 36

    approximate formula 68

    formula 67

    formula versus friction
        factor 67

    nozzle steam formula 109

velocity regime 25

viscosity 64

    formula 64

    of air 64

    reference temperature 64

    Sutherland temperature 64

volume flow 13

W

Wantzel Pierre 44

water pumps 4

wellhead 27

working substance 41

Z

Zeuner formula of k 47

www.ingramcontent.com/pod-product-compliance
Lightning Source LLC
Chambersburg PA
CBHW030918180526
45163CB00002B/380